Elementary Number Theory

Jie Xiao

International Press
www.intlpress.com

Elementary Number Theory
by Jie Xiao (Memorial University of Newfoundland)

ISBN 978-1-57146-183-4

Paperback reissue 2010.
Previously published in 2006 under ISBN 978-1-57146-163-6 (clothbound).

Typeset using the LaTeX system.
[Printed in the USA on acid-free paper].

Preface

Traditionally, elementary number theory is a branch of number theory dealing with the integers without use of techniques from other mathematical fields. With this objective in mind, and exercising as much control as possible over my own prejudices, I have sought to pare away all material that might be considered extraneous in a three hour per week, twelve week semester course in elementary number theory. This leads to my aim in writing this book: On the one hand, I must present in a well-motivated and natural sequence the basic ideas and results of elementary number theory. On the other hand, enough material is covered to provide a firm base on which to build for later studies in algebraic number theory and analytic number theory. The only background material required of the reader is a knowledge of some simple properties of the system of integers. Otherwise this concise book is self-contained.

The book begins with a few preliminaries on induction principles, followed by a quick review of division algorithm. The substance of the book starts in the second chapter, where, using divisors, the greatest (least) common divisor (multiple), the Euclidean algorithm and linear indeterminate equation are discussed. This foundation supports the subsequent chapters: prime numbers; congruences; congruent equations; and, finally, three additional topics (comprising cryptography, Diophantine equations and Gaussian integers). Placed at the end of each chapter are some exercises that illustrate the theory and provide practice in the techniques. Answers to all the even-numbered problems are given at the end of the book.

The above-mentioned material was used with groups of undergraduates in one-semester courses at Memorial University of Newfoundland. A brisk pace should make it possible to cover this little book in its entirety in one semester.

While writing this book I was encouraged by and benefited from a number of individuals. Here, I want to thank Hershy Kisilevsky at Concordia University as well as Herb Gaskill, Donald E. Rideout, Yiqiang Zhou and my students at Memorial University of Newfoundland for their helpful comments and suggestions. Meanwhile, I am very grateful to the referee for his invaluable review, but also to Shing-Tung Yau at Harvard University and Brian Bianchini and Lisa Lin at International Press for their practical advice. Last but not least, I wish to take this opportunity to acknowledge my sense of indebtedness to my family for their considerable patience and understanding.

Montreal & St. John's
Summer 2002 *& Fall* 2005 – *Spring* 2006 *Jie Xiao*

Contents

CHAPTER 1

Basics

In this chapter we review two basic methods of mathematical proof – mathematical induction and contradiction argument, as well as the division algorithm.

1.1. Principles of Induction and Well-ordering

From now on, let **N**, **Z**, **Q** and **R** be the set of natural numbers, integers, rational numbers, and reals. Like the rest of mathematics, **N** is a postulational system. For example, we might take the integers themselves and the operations of addition and multiplication as undefined terms and statements such as the closure law for addition, which states that the sum of any two elements in **N** is in **N**, as postulates. In general, these postulates are well understood and we will not consider them here. However, we do consider one postulate that is often found confusing. The postulate in question is the *Principle of Mathematical Induction* (*PMI*) or its equivalent, the *Well-ordering Principle* (*WP*). Below are three alternative statements (typically used to establish that a given statement is true of all natural numbers, or otherwise is true of all members of an infinite sequence):

I_1. *First form of the PMI.* Any set of positive integers that contains 1, and that contains $k+1$ whenever it contains k, contains all positive integers.

I_2. *Second form of the PMI.* Any set of positive integers that contains 1, and that contains $k+1$ whenever it contains positive integers 1 to k inclusive, contains all positive integers.

I_3. *The WP.* Every nonempty set of positive integers contains a least element.

To better understand the above two principles, we illustrate the use of them by assuming them as postulates and using them to prove three results.

Example 1.1.1. For $n \in \mathbf{N}$, $\sum_{i=1}^{n} i = n(n+1)/2$.

Proof. Let S be the set of all $n \in \mathbf{N}$ for which the formula of the example is true. Clearly, $1 \in S$ since for $n = 1$, the assertion is simply that $1 = (1+1)/2$. Now suppose that $k \in S$, namely,

$$\sum_{i=1}^{k} i = k(k+1)/2.$$

Then

$$\begin{aligned}\sum_{i=1}^{k+1} i &= \sum_{i=1}^{k} i + (k+1)\\ &= k(k+1)/2 + (k+1)\\ &= (k+1)(k+2)/2.\end{aligned}$$

Hence $k+1 \in S$. Therefore, it follows from I_1 that S contains $\mathbf{N}$, i.e., the given formula is valid for $\mathbf{N}$, as claimed. □

A proof based on I_2 is exactly like one based on I_1, with one exception. In the second step of the proof, one assumes the truth of the assertion for all values of n from 1 to k inclusive and, on the basis of this assumption, must then prove its truth for $n = k+1$. The point is that the truth of the $(k+1)$th case often does not follow directly from the truth of the kth case, but does follow from the truth of the assertion for some or all of the positive integers preceding $k+1$. Even in such cases, it is possible to use I_1 by a devious trick, but a proof based on I_2 would be much more natural.

Example 1.1.2. For $n \in \mathbf{N}$ let a_n be the nth term of the Fibonacci sequence which is defined by the following conditions:

$$a_1 = 1 = a_2, \quad a_n = a_{n-1} + a_{n-2}, \quad n \geq 3.$$

Then $a_{n+1} < (7/4)^n$ for any $n \in \mathbf{N}$.

Proof. If $n = 1$ then $a_2 = 1 < 7/4$ and hence the inequality holds for $n = 1$. Assume that it is true for all integers n: $1 \leq n \leq k$. Then for $n = k+1$ one has

$$\begin{aligned}a_{k+2} &= a_{k+1} + a_k\\ &< (7/4)^k + (7/4)^{k-1}\\ &= (7/4)^{k-1}(7/4+1)\\ &< (7/4)^{k+1}.\end{aligned}$$

Thus the inequality is true for $k+1$ and hence for all integers $n \geq 1$ by I_2. □

In fact, this argument suggests an alternative *PMI* as follows.

I_4. *Third form of the PMI.* Any set of positive integers that contains 1 and 2, and that contains $k+2$ whenever it contains positive integers k to $k+1$, contains all positive integers.

The third example of proof based on the well-ordering principle is Hugo Steinhaus's argument for $\sqrt{2} \in \mathbf{R} \setminus \mathbf{Q}$.

Example 1.1.3. The number $\sqrt{2}$ is irrational.

Proof. Since $1 < 2 < 4$, it follows that $1 < \sqrt{2} < 2$. Now suppose $\sqrt{2}$ belongs to $\mathbf{Q}$. Then by I_3, there exists a least positive integer b and an $a \in \mathbf{N}$

such that $\sqrt{2} = a/b$. This implies $1 < a/b < 2$ and hence $0 < a - b < b$. However, $2 = a^2/b^2$ yields $2b^2 - ab = a^2 - ab$ and so

$$\sqrt{2} = \frac{a}{b} = \frac{2b - a}{a - b}.$$

This gives $\sqrt{2}$ as a ratio of two integers with a positive integer denominator less than b. Since this is a contradiction, the result is true. □

As in this case, many proofs based on the *WP* involve the method of proof by contradiction. To prove a result by contradiction, one proceeds, in general, as follows. One begins by assuming that the result is false, and then deduces from this assumption a result that is known to be false, or that contradicts the primary assumption. Of course, this is only one of the indirect arguments. Another one is the *contrapositive argument* which is made by assuming that the result B is false and deducing that the condition A, too, must be false, whenever the implication is: *if A then B, written* $A \Rightarrow B$.

1.2. Equivalence of Principles

In this section we show that I_1, I_2 and I_3 are equivalent. The proof does not use the standard mathematical symbols for *there exists* and *for all* to make it more accessible to less mathematically motivated readers. The key technique is natural deduction logic and proof by contradiction.

First, we need a lemma as follows.

Lemma 1.2.1. *If $a \in \mathbf{N}$ then $a \geq 1$, that is, 1 is the least positive integer.*

PROOF.

Method 1. It is clear that $1 \geq 1$. Assume that $k \geq 1$ where k is any fixed but unspecified positive integer. Then $k + 1 \geq k \geq 1$ and it follows by I_1 that $n \geq 1$ for every $n \in \mathbf{N}$. Thus I_1 implies that 1 is the least positive integer.

Method 2. I_2 may be applied to prove the result. The argument is similar.

Method 3. We use I_3 to prove the result. Suppose, on the contrary, that there is an integer $a \in (0, 1)$. Then, if S is the set of all such integers, it is not empty. Therefore, by I_3, S must have a least element. Let b be the least element of S. Then $0 < b < 1$, consequently $0 < b^2 < b < 1$ and $b^2 \in S$. However, this contradicts the fact that b was the least one in S. Because of this contradiction, our original assumption must be false, so that $a \geq 1$ for every $a \in \mathbf{N}$. □

As a consequence of Lemma 1.2.1, we have that if $k \in \mathbf{N}$ then there is no $n \in \mathbf{N}$ such that $k < n < k+1$. To see this, note that if there exists such $n \in \mathbf{N}$ that $k < n < k + 1$ then $0 < n - k < 1$, so that $n - k$ is a positive integer less than 1, in contradiction to Lemma 1.2.1. Thus no such n can exist.

Theorem 1.2.1. *I_1, I_2 and I_3 are equivalent.*

PROOF. $I_1 \Rightarrow I_2$. We take I_1 as a postulate and prove I_2 as a theorem. Suppose S is any set of positive integers satisfying the conditions of I_2. The problem is to show that S contains $\mathbf{N}$.

Let A_n denote the statement: the integers 1 to n inclusive are in S. A_1 is true by hypothesis. Now, assume that A_k is true. Then 1 to k inclusive are in S. Hence, again by hypothesis, $k+1$ is in S and A_{k+1} is true. Therefore, by I_1, A_n is true for any $n \in \mathbf{N}$ with $n \geq 1$, so S contains $\mathbf{N}$.

$I_2 \Rightarrow I_3$. We take I_2 as a postulate and prove I_3 as a theorem. Suppose S is any nonempty set of positive integers. The problem is to show that S has a least element.

Assume that S has no least element and let A_n denote the statement: n is not an element of S. Then A_1 is true, for otherwise 1 would be the least element of S by Lemma 1.2.1, which follows from I_2. Assume that A_n is true for all n from 1 to k inclusive. Then, A_{k+1} must also be true, for otherwise $k+1$ would be the least element in S. Thus, by I_2, A_n is true for all $n \in \mathbf{N}$ with $n \geq 1$. But this implies that S is an empty set, contrary to the hypothesis. Therefore, the assumption that S has no least element is false and the desired implication is proved.

$I_3 \Rightarrow I_1$. Assuming I_3 as a postulate, we must verify that S contains all $n \in \mathbf{N}$, where S is a set of positive integers obeying the conditions of I_1.

If S does not contain all $n \in \mathbf{N}$ with $n \geq 1$, then the set S' of all $n \notin S$ is nonempty. Therefore, by I_3, S' has a least element. It follows from Lemma 1.2.1 that no elements of S' are less than 1 and hence that the least element of S' is not less than 1. Thus, the least element in S' can be written in the form $k+1$. But this says that k is in S whereas $k+1$ is not, in direct contradiction to the hypothesis. Thus, the assumption that S does not contain all positive integers ≥ 1 is false and the theorem is proved. □

1.3. Division Algorithm

The *WP* may be modified to give the following division algorithm – one of the most fundamental results in number theory.

Theorem 1.3.1. *If a and b are integers with $b \geq 1$, there exist unique integers q and r such that $a = qb + r$, $0 \leq r < b$.*

PROOF. Existence. Suppose S is the set of all integers of the form $a - sb \geq 0$. If $a \geq 0$, then $a - 0 \cdot b \in S$. If $a < 0$, then

$$0 \leq a(1-b) = a - ab \in S.$$

Thus, S is not empty in either case. Hence, by the *WP*, S has a least element. Let q denote the value of s which yields the least element $r = a - bq \in S$.

Thus,

$$r \geq 0 \quad \text{and} \quad r - b = a - bq - b = a - (q+1)b < 0$$

due to $r - b$ being of the form $a - sb$ and yet being less than the least nonnegative integer of this form. So, $0 \leq r < b$, as claimed.

Uniqueness. To show that q and r are unique, we suppose that $a = bq' + r'$, where $q', r' \in \mathbf{Z}$ and $0 \leq r' < b$. If suffices to verify that $r = r'$ and $q = q'$. If $q' < q$, then $q' + 1 \leq q$ and hence

$$r = a - bq \leq a - b(q' + 1) = r' - b < 0,$$

which is a contradiction. Similarly, we obtain a contradiction if $q' > q$. Therefore, it must be the case that $q' = q$ and so $r' = r$. □

Note that the restriction $b \in \mathbf{N}$ is not strictly necessary, and the theorem could also be written in the following form. *Given $a, b \in \mathbf{Z}$ with $b \neq 0$, there exist unique $q, r \in \mathbf{Z}$ with $0 \leq r < |b|$ such that $a = bq + r$. As usual, r is called the remainder of a.*

Example 1.3.1. For the four cases with $a = \pm 275$ and $b = \pm 61$ we have

$$\begin{aligned} 275 &= 4 \cdot 61 + 31 \\ 275 &= (-4) \cdot (-61) + 31 \\ -275 &= (-5) \cdot 61 + 30 \\ -275 &= 5 \cdot (-61) + 30. \end{aligned}$$

The division algorithm is very useful. For instance, if $b = 2$ then the theorem implies that every $a \in \mathbf{Z}$ is either of the form $2k$ (even) or of the form $2k+1$ (odd) (consequently, a^2 is either of the form $4k^2$ or $4(k^2+k)+1$). Similarly, any $a \in \mathbf{Z}$ must be of the form $3k$, or $3k+1$, or $3k+2$ (consequently, a^2 is either of the form $9k^2 = 3u$ or $9k^2 + 6k + 1 = 3v + 1$, or $9k^2 + 12k + 4 = 3w + 1$). Admittedly, these are only small results, but they are not without interest and they indicate an important way in which the division algorithm can be used.

Exercises

1.1. Show that every odd natural number can be written in the form $4n + 1$ or $4n + 3$.

1.2. Prove $\sum_{i=1}^{n} 1/(i^2 + i) = n/(n+1)$ for $n \in \mathbf{N}$.

1.3. Prove $\sum_{i=1}^{n} i^2 = n(n+1)(2n+1)/6$ for $n \in \mathbf{N}$.

1.4. Prove $\sum_{i=1}^{n} i^3 = (n(n+1)/2)^2$ for $n \in \mathbf{N}$.

1.5. Prove $(2n)! < 2^{2n}(n!)^2$ for $n \in \mathbf{N}$.

1.6. Prove that for $x \neq -1$ and $n \in \mathbf{N}$,

$$x^{2n} - x^{2n-1} + \cdots + x^2 - x + 1 = \frac{x^{2n+1} + 1}{x + 1}.$$

1.7. Use the mathematical induction to prove that 16^n always ends in a 6 for any $n \in \mathbf{N}$.

1.8. Use I_4 to prove that

$$\left(\frac{1+\sqrt{5}}{2}\right)^{n-2} < \frac{\left(\frac{1+\sqrt{5}}{2}\right)^n - \left(\frac{1-\sqrt{5}}{2}\right)^n}{\sqrt{5}},$$

for any $n \in \mathbf{N}$ with $n \geq 3$.

1.9. Use the well-ordering principle to verify the Archimedean axiom: if $m, n \in \mathbf{N}$ then there is a $k \in \mathbf{N}$ such that $kn \geq m$.

1.10. Use the well-ordering principle to prove that $\sqrt{5}$ is irrational.

1.11. Prove that the product of any two odd numbers must be odd.

1.12. Prove that if a and b are integers with $b < 0$ then there exist unique integers q and r with $0 \leq r < |b|$ such that $a = bq + r$.

CHAPTER 2

Divisibility

In this chapter, we introduce the concept of divisibility of integers, which is viewed as one of the most important ideas in number theory, and consider some further questions regarding divisibility.

2.1. Divisors

Let us begin with the following definition.

Definition 2.1.1. If $a, b \in \mathbf{Z}$ with $a \neq 0$ and there exists $c \in \mathbf{Z}$ such that $b = ac$, then we say that a divides b and write $a|b$. We also call a a divisor of b and b a multiple of a. If $1 \leq a < b$ and $a|b$, then a is called a proper divisor of b. If a does not divide b, then we write $a \nmid b$.

We now state a number of simple facts involving the concept of divisor as follows.

Theorem 2.1.1. *Let $a, b, c \in \mathbf{Z}$ with $a \neq 0$. Then*

(i) *$a|b$ implies $a|bc$.*
(ii) *$a|b$ and $b|c$ imply $a|c$.*
(iii) *$a|b$ and $a|c$ imply $a|(bx + cy)$ for $x, y \in \mathbf{Z}$.*
(iv) *$a|b$ and $b|a$ imply $a = \pm b$.*
(v) *$a|b$ and $b \neq 0$ imply $|a| \leq |b|$.*
(vi) *For $m \in \mathbf{Z}$ with $m \neq 0$, $a|b$ if and only if $ma|mb$.*

PROOF. It suffices to verify (v). Since $a|b$, there is $c \in \mathbf{Z}$ such that $b = ac$. This gives $|b| = |a||c| \geq |a|$. □

It is worth mentioning that (iii) may extend to sums of more than two terms as follows. *If $a|b_i$ for $b_1, b_2, \ldots, b_n \in \mathbf{Z}$, then $a|\sum_{i=1}^{n} b_i x_i$ for $x_1, x_2, \ldots, x_n \in \mathbf{Z}$.*

2.2. The Greatest Common Divisor

For $a, b \in \mathbf{Z}$, if $d|a$ and $d|b$, then d is said to be a common divisor of a and b. If $a = b = 0$, then it is clear that they have infinitely many common divisors. However, if at least one of them is not zero, then (v) produces that the number of common divisors is finite and hence that there must be a largest one.

Definition 2.2.1. Let $a, b \in \mathbf{Z}$, not both zero. Then $g \in \mathbf{Z}$ is called the greatest common divisor (*gcd*) of a and b, and is denoted by $g = (a, b)$, provided the following three conditions hold:

(i) $g|a$ and $g|b$.
(ii) If $d|a$ and $d|b$, then $d|g$.
(iii) $g > 0$.

Observe that if one of a and b is small, then the problem of finding (a, b) is not difficult since there are only a few alternatives. For example, it is easy to see that ± 1, ± 2, ± 3 and ± 6 are the only common divisors of 12 and 18 and that $6 = (12, 18)$. However, trial-and-error methods are not very efficient when it comes to large values of a and b. There is an efficient and systematic way for finding (a, b), but before discussing it, it will be convenient to present an interesting and very useful alternative characterization of the *gcd*.

Theorem 2.2.1. *Let $a, b \in \mathbf{Z}$, not both zero. Then $g = (a, b)$ is the least element of $S = \{ax + by \in \mathbf{N} : \ x, y \in \mathbf{Z}\}$.*

PROOF. By the hypothesis, at least one of a and b is not zero. For definiteness, suppose $a \neq 0$. If $a > 0$ then $a \in S$, and if $a < 0$ then $-a \in S$. Thus, S is not empty, and so, by the *WP*, S must have a least element, say,

$$s = ax_0 + by_0 \quad \text{where} \quad x_0, y_0 \in \mathbf{Z}.$$

Accordingly, it suffices to show $s = g$.

By the division algorithm there are $q, r \in \mathbf{Z}$ with $0 \leq r < s$ such that $a = sq + r$. Thus,

$$r = a - sq = a - (ax_0 + by_0)q = a(1 - qx_0) + b(-qy_0) \in S.$$

If r were not zero, then it would be the least element of S, contradicting our assumption that s is the smallest one. Thus, $a = sq$ and $s|a$. Similarly, one has $s|b$. So s is a common divisor of a and b, so that by the definition, $s \leq g$. On the other hand, since $s = ax_0 + by_0$ and $g|a$ and $g|b$, it follows from Theorem 2.1.1 (iii) that $g|s$ and so that $g \leq s$ thanks to Theorem 2.1.1 (v). Therefore $s = g$. □

This theorem tells us that the *gcd* is unique.

Corollary 2.2.1. *Let $a, b \in \mathbf{Z}$, not both zero. Then*

(i) *For $m \in \mathbf{N}$, $(ma, mb) = m(a, b)$.*
(ii) *$d|a$ and $d|b$ with $d > 0$ imply $(a/d, b/d) = (a, b)/d$.*
(iii) *$(a, b) = g$ implies $(a/g, b/g) = 1$.*

PROOF. It suffices to verify (i). By Theorem 2.2.1 we obtain that (ma, mb) is the least positive value of $max + mby$ which equals m times the least positive value of $ax + by$. So the result follows. □

2.3. The Euclidean Algorithm

We are now in a position to discuss an orderly and systematic process for finding the *gcd* of two nonzero integers.

Lemma 2.3.1. *Let* $a, b \in \mathbf{N}$. *Then* $(a, b) = (a, b - ax)$ *for any* $x \in \mathbf{Z}$.

PROOF. Assume $g = (a, b)$ and $h = (a, b - ax)$. By Theorem 2.1.1 (iii) $g|(b-ax)$. Hence g is a common divisor of a and $b-ax$ and so $g \leq h$. Again by Theorem 2.1.1 (iii) we have $h|\big(ax + (b - ax)1\big)$, that is, $h|b$ since $h|a$ and $h|(b-ax)$. Thus h is a common divisor of a and b and therefore $h \leq g$. This gives $g = h$. $\square$

Lemma 2.3.1 indicates that given two integers, not both zero, subtracting a multiple of one number from the other changes the pair but not the *gcd*. For example,

$$\begin{aligned}
(210, 679) &= (210, 679 - 3 \cdot 210) \\
&= (210, 49) \\
&= (210 - 4 \cdot 49, 49) \\
&= (14, 49) \\
&= (14, 49 - 3 \cdot 14) \\
&= (14, 7) \\
&= (14 - 2 \cdot 7, 7) \\
&= (0, 7) = 7.
\end{aligned}$$

As we saw above, Lemma 2.3.1 and the division algorithm can be used to compute the *gcd* of two integers effectively. We can also calculate the values of x_0 and y_0 such that $g = ax_0 + by_0$. Considering the above example again $(a = 679, b = 210)$, we have by the division algorithm:

$$\begin{aligned}
679 &= 3 \cdot 210 + 49 \\
210 &= 4 \cdot 49 + 14 \\
49 &= 3 \cdot 14 + 7 \\
14 &= 2 \cdot 7 + 0.
\end{aligned}$$

We may express 7 as a linear combination of 679 and 210 by eliminating the remainders 14 and 49 as follows:

$$\begin{aligned}
7 &= 49 - 3 \cdot 14 \\
&= 49 - 3 \cdot (210 - 4 \cdot 49) \\
&= 13 \cdot 49 - 3 \cdot 210 \\
&= 13 \cdot (679 - 3 \cdot 210) - 3 \cdot 210 \\
&= 13 \cdot 679 + (-42) \cdot 210.
\end{aligned}$$

So, $x_0 = 13$ and $y_0 = -42$. Of course, these values of x_0 and y_0 are not unique. For instance,

$$x_0 = 13 + 210k \quad \text{and} \quad y_0 = -42 - 679k \quad \text{where} \quad k \in \mathbf{Z},$$

will do the job. This process suggests a general method of finding the *gcd* of two positive integers, that is, the Euclidean Algorithm as follows:

Theorem 2.3.1. *Given $a, b \in \mathbf{N}$, we have by the division algorithm the following equations:*

$$\begin{aligned} a &= bq_1 + r_1, \quad 0 < r_1 < b, \\ b &= r_1q_2 + r_2, \quad 0 < r_2 < r_1, \\ r_1 &= r_2q_3 + r_3, \quad 0 < r_3 < r_2, \\ \dots &= \dots, \quad \dots, \\ r_{j-2} &= r_{j-1}q_j + r_j, \quad 0 < r_j < r_{j-1}, \\ r_{j-1} &= r_jq_{j+1} + 0. \end{aligned}$$

The gcd of a and b is r_j, the last nonzero remainder in the division process. Values of x_0 and y_0 in $(a, b) = ax_0 + by_0$ can be obtained by eliminating the remainders $r_{j-1}, \dots, r_2, r_1$ from the system of equations.

PROOF. The chain of equations must eventually give remainder 0 by the *WP*. We assume that the $(j+1)$th equation is the first occurrence of the remainder 0. By Lemma 2.3.1, we have

$$\begin{aligned} (a, b) &= (a - bq_1, b) \\ &= (r_1, b) \\ &= (r_1, b - r_1q_2) \\ &= (r_1, r_2) = \dots \\ &= (r_{j-1} - r_jq_{j+1}, r_j) \\ &= (0, r_j) = r_j. \end{aligned}$$

To see that r_j is expressible as a linear combination of a and b, we need to merely eliminate $r_{j-1}, \dots, r_1$ starting from the equation $r_j = r_{j-2} - r_{j-1}q_j$ and working backwards through the equations. □

2.4. Linear Indeterminate Equation

To study the problem of solving the linear indeterminate equation $ax + by = n$, we need a simple but important lemma.

Lemma 2.4.1. *Let $a, b, c \in \mathbf{Z}$. If $c|ab$ and $(b, c) = 1$, then $c|a$.*

PROOF. Since $(b, c) = 1$, Theorem 2.2.1 is applied to imply that there are $x, y \in \mathbf{Z}$ such that $bx + cy = 1$, and then $abx + acy = a$. Since $c|ab$ and $c|ac$, $c|a$. □

Lemma 2.4.1 is employed to give all integer solutions of the equation $ax + by = n$.

Theorem 2.4.1. *If $a, b, n \in \mathbf{Z}$ with $(a, b) = 1$, then $ax + by = n$ has a solution $x = x_0$ and $y = y_0$ in $\mathbf{Z}$. Moreover, all integer solutions are given by $x = x_0 + bt$ and $y = y_0 - at$, $t \in \mathbf{Z}$.*

PROOF. Since $(a, b) = 1$, by Theorem 2.2.1 there are $x_1, y_1 \in \mathbf{Z}$ such that $ax_1 + by_1 = 1$. Then $x_0 = nx_1$ and $y_0 = ny_1$ is a solution of $ax + by = n$. Let x' and y' be another solution, then

$$a(x_0 - x') + b(y_0 - y') = 0 \quad \text{and hence} \quad a(x_0 - x') = -b(y_0 - y').$$

By $(a, b) = 1$, we may assume $b \neq 0$ (otherwise $a = 0 = b$ implies $(a, b) \neq 1$). It follows from Lemma 2.4.1 that $b|(x_0 - x')$ and so that $x_0 - x' = -bt$ for some $t \in \mathbf{Z}$, or $x' = x_0 + bt$. Consequently, $abt = b(y_0 - y')$, which yields $y' = y_0 - at$.

The fact that the integers $x = x_0 + bt$ and $y = y_0 - at$ satisfy $ax + by = n$ for all $t \in \mathbf{Z}$ can be checked by substituting them in the given equation. □

Generally speaking, we have the following assertion. *Let $(a, b) = g$. Then $ax + by = n$ has an integer solution if and only if $g|n$.* The necessity is trivial. For the sufficiency, if $g|n$ then $ax + by = n$ and $(a/g)x + (b/g)y = n/g$ have the same solutions, and hence Theorem 2.4.1 covers the general situation since g can be divided out.

As a numerical example, let us consider the following equation.

Example 2.4.1. Solve all integers x and y for $355x + 34y = 3$.

Solution. It is clear that

$$(355, 34) = 1 \quad \text{and} \quad 355 \cdot (-9) + 34 \cdot 94 = 1.$$

Thus $x_0 = -27$ and $y_0 = 282$ is a solution of the equation. By Theorem 2.4.1 we see that all integer solutions of the equation are determined by

$$x = -27 + 34t \quad \text{and} \quad y = 282 - 355t, \quad t \in \mathbf{Z}.$$

Below is an example to show that not all linear indeterminate equations have positive integer solutions.

Example 2.4.2. Solve all $x, y \in \mathbf{Z}$ for $8393x + 3437y = 301$.

Solution. Although it is not easy to see what $(8393, 3437)$ is, we may apply the Euclidean Algorithm to find

$$(8393, 3437) = 7 \quad \text{and} \quad 7 = 8393 \cdot (-224) + 3437 \cdot (547)$$

and hence

$$1 = 1199 \cdot (-224) + 491 \cdot (547).$$

Therefore, the equation is equivalent to $1199x + 491y = 43$. One of its special solutions is

$$x_0 = -224 \cdot 43 = -9632 \quad \text{and} \quad y_0 = 547 \cdot 43 = 23521$$

and all solutions are given by

$$x = -9632 + 491t \quad \text{and} \quad y = 23521 - 1199t, \quad t \in \mathbf{Z}.$$

The positive solutions are found by solving simultaneously the following inequalities:

$$-9632 + 491t > 0 \quad \text{and} \quad 23521 - 1199t > 0.$$

However, there is no such $t \in \mathbf{Z}$ satisfying both inequalities, and so the equation has no positive integer solutions.

2.5. The Least Common Multiple

If $a|m$ and $b|m$, then m is called a common multiple of integers a and b. Since division by zero is meaningless, it is clear that this definition has meaning only if a and b are both different from zero. In this case it is obvious that ab and $-ab$ are both common multiples of a and b and that one of them is positive. Therefore, by the *WP*, there must exist a least positive common multiple.

Compared to the *gcd*, the concept of least common multiple can be introduced below.

Definition 2.5.1. The least common multiple (*lcm*) of nonzero integers a and b, written $[a, b]$, is the unique positive integer m with the following two properties:

(i) $a|m$ and $b|m$.
(ii) If $a|c$ and $b|c$, then $m|c$.

It is clear that there cannot exist two different positive integers having (i) and (ii) of the above definition. The existence of *lcm* will be an immediate consequence of the following result.

Theorem 2.5.1. *Let a and b be nonzero integers. Then*

(i) $[ma, mb] = m[a, b]$ *for any* $m \in \mathbf{N}$.
(ii) $[a, b] = |ab|/(a, b)$.

PROOF.

(i) Let $H = [ma, mb]$ and $h = [a, b]$. Then mh is multiple of ma and mb and hence $H \le mh$. Also, H is a multiple of both ma and mb, and so H/m is a multiple of a and b. Thus $h \le H/m$, that is, $mh \le H$, from which it follows that $mh = H$.

(ii) It suffices to check the formula for positive integers a and b.

Special case: $(a, b) = 1$. With this, we have to show $[a, b] = ab$. Since $[a, b]$ is a multiple of a, we may let $[a, b] = na$. Then $b|na$. Note that $(a, b) = 1$. So Lemma 2.4.1 implies $b|n$ and so $n = bs$ for some $s \in \mathbf{Z}$. Hence $[a, b] = abs$, and clearly s must be 1 since abs is the least positive common multiple of a and b.

General case: $(a, b) = g$. With this, we have $(a/g, b/g) = 1$ and use the previous paragraph to obtain

$$\left[a/g, b/g\right]\left(a/g, b/g\right) = (a/g) \cdot (b/g).$$

Multiplying by g^2 and using Corollary 2.2.1 (i) as well as (i) of the present theorem, we reach $[a, b](a, b) = ab$. □

Of course, Theorem 2.5.1 gives a relationship between *gcd* and *lcm*. For example, $(8393, 3437) = 7$ yields

$$[8393, 3437] = (8393 \cdot 3437)/7 = 4120963.$$

So far we have considered the *gcd* and the *lcm* of two integers, but it is fairly easy to extend the concept and the fundamental properties to more than two integers. We leave the details as exercises.

Exercises

2.1. Prove that if $k|(35n + 26)$ and $k|(7n + 3)$ for $k \in \mathbf{N}$ then $k = 1$ or 11.

2.2. Prove that if integers a, b, q, r obey $a = bq + r$ with $0 \leq r < b$ and $b|a$ then $r = 0$.

2.3. Prove that if positive integers k, l, m, n, p, q satisfy

$$(k, l) = p, \ (k, m) = q \quad \text{and} \quad (l, m) = 1,$$

then $(k, lm) = pq$. Give an example to show that this need not be true if $(l, m) > 1$.

2.4. Compute $(357, 629)$ and determine integers x and y such that $(357, 629) = 357x + 629y$.

2.5. Show that if $a, b, c, d, x, y \in \mathbf{N}$ satisfy

$$ax + by = m, \ cx + dy = n \text{ and } ad - bc = \pm 1,$$

then $(m, n) = (x, y)$.

2.6. Prove that given $a, b \in \mathbf{N}$, then $b|a$ if and only if $(a, b) = |b|$.

2.7. Prove that if $m, n \in \mathbf{N}$ then

$$(a^m - 1, a^n - 1) = a^{(m,n)} - 1 \quad \text{for any} \quad a - 1 \in \mathbf{N}.$$

2.8. Given $a, b \in \mathbf{N}$ with $(a, b) = 1$, prove

$$(a + b, a - b) = 1 \quad \text{or} \quad (a + b, a - b) = 2.$$

2.9. Find all possible integers x and y such that

(i) $28x + 35y = 60$;
(ii) $343x + 259y = 658$;
(iii) $12x + 57y = 423$.

2.10. Find $[357, 629]$ and $[1492, 1066]$.

2.11. Define $k\mathbf{Z} = \{kz : z \in \mathbf{Z}\}$ for a given $k \in \mathbf{N}$. Prove that if $m, n \in \mathbf{N}$ then $[m, n]\mathbf{Z} = m\mathbf{Z} \cap n\mathbf{Z}$.

2.12. Give the definitions of the *gcd*: $(a_1, \ldots, a_n)$ and the *lcm*: $[a_1, \ldots, a_n]$ of n positive integers: $a_1, \ldots, a_n$. Prove that if none of $a_1, \ldots, a_n$ is zero, then

$$(a_1, \ldots, a_n) = ((a_1, \ldots, a_{n-1}), a_n)$$

and

$$[a_1, \ldots, a_n] = [[a_1, \ldots, a_{n-1}], a_n].$$

CHAPTER 3

Primes

In this chapter, we are devoted to the study of prime numbers beginning with the problem of determining whether a given integer is prime or composite. Although this problem has no simple general solution, it is possible to give partial results including the Fundamental Theorem of Arithmetic.

3.1. Number of Primes

We start with the definition of a prime.

Definition 3.1.1. A positive integer $p \neq 1$ is said to be a prime if its positive divisors are 1 and p. If $p > 1$ is not a prime (i.e., there are $s, t \in \mathbf{N}$ with $s, t < p$ such that $p = st$), then it is called composite.

For instance, $2, 3$ and 5 are primes, but 6 is composite. Of course, 1 is neither prime nor composite.

Lemma 3.1.1. *Every integer $n > 1$ has a prime factor.*

PROOF. Let S denote the set all positive divisors of n that are greater than 1. Since $n \in S$, the set S is not empty and it thus contains a least element p thanks to I_3. Now any divisor of p is also a divisor of n and if p were composite, S would contain smaller elements than p. Accordingly, p is prime and the lemma is proved. □

The following result is due to Euclid of Alexandria and so has been known for more than 2000 years.

Theorem 3.1.1. *The number of all primes is infinite.*

PROOF. Suppose there are only r primes, say $p_1, \ldots, p_r$, and consider $n = p_1 \cdots p_r + 1$. By Lemma 3.1.1, n has a prime factor p. Moreover, no p_i is a factor of n, for otherwise it would follow that $p_i | 1$. Accordingly, there are $r + 1$ primes: $p_1, \ldots, p_r, p$. This contradicts the previous assumption. □

There are various simple and useful ways of classifying the primes. First, let us observe that any $n \in \mathbf{N}$ can be written as either $2k$ or $2k+1$ for some nonnegative integer k. Now $2k$ cannot be prime unless $k = 1$, so every prime except 2 must be odd. By the last theorem, the number of odd primes is therefore infinite. Moreover, Johann Peter Gustav Lejeune Dirichlet proved in 1837 that *if $a, b \in \mathbf{N}$ are such that $(a, b) = 1$, then there exist infinitely many primes of the form $a + kb$, $k \in \mathbf{N}$.* It is clear that the conditions of

Dirichlet's result are necessary since if $(a,b)=g>1$, then $g|(a+kb)$ for any $k\in\mathbf{N}$ and $a+kb$ is never a prime. The difficulty arises in showing the conditions are sufficient, but no simple proof of this direction is known except certain special cases.

3.2. The Fundamental Theorem of Arithmetic

Before considering the Fundamental Theorem of Arithmetic, we need the following lemma.

Lemma 3.2.1. *Let p be a prime and $a,b\in\mathbf{N}$. If $p|ab$ then $p|a$ or $p|b$. More generally, if $p|a_1a_2\cdots a_n$ for $a_1,\ldots,a_n\in\mathbf{N}$ then $p|a_i$ for some i with $1\le i\le n$.*

PROOF. Since the second part can easily be proved by induction, it is enough to prove the first part. If $p|a$ then the result follows. Otherwise, if $p\not|a$ then $(p,a)=1$ since p is prime. By Theorem 2.2.1 , there are $x,y\in\mathbf{Z}$ such that $px+ay=1$. Multiplying this equation by b we have $pbx+aby=b$. Clearly p divides the left-hand-side of the last equation and so $p|b$. □

The following is the fundamental theorem mentioned above.

Theorem 3.2.1. *Every integer $n\ge 2$ is either a prime or can be represented as a product of primes. This factoring is unique except for the order of the prime factors.*

PROOF. We show the first part by induction. The assertion is trivially true for $n=2$. Assume that it is true for all integers n with $2\le n\le k$. We must show that the assertion is also true for $k+1$. If $k+1$ is prime, there is nothing to show. If $k+1$ is not prime, there exist $r,s\in\mathbf{N}$ with $2\le r,s\le k$ such that $k+1=rs$. By the assumption, r,s are either primes or products of primes. Thus, $k+1$ must be a product of at least two primes. Therefore, I_2 implies the assertion is true for all $n\in\mathbf{N}$ with $n\ge 2$.

The proof of the second part needs Lemma 3.2.1. Assume that there is an $n\in\mathbf{N}$, $n\ge 2$, with two different factorizations:

$$n=p_1\cdots p_r=q_1\cdots q_s,$$

where the p's and q's are primes and the p's are not identical with the q's. Dividing out any primes common to the two representations, we would have an equality which, by a suitable choice of notation, may be written in the form

$$p_1\cdots p_i=q_1\cdots q_j.$$

However, this is impossible because $p_1|q_1\cdots q_i$ and by Lemma 3.2.1, $p_1|q_k$ for some k. That is to say, p_1 must be identical with at least one of the q's. This contradiction indicates that our assumption is wrong, and the proof of the theorem is completed. □

In expressing a given integer $n > 1$ as a product of primes, it is often useful to adopt a notation which exhibits all the distinct prime factors of n. The preceding theorem assures us that n can be uniquely written as $p_1^{k_1} \cdots p_r^{k_r}$, where $p_1, \ldots, p_r$ are the distinct primes factors of n, and $k_1, \ldots, k_r \in \mathbf{N}$. *Such a factorization is called the canonical/standard form/factorization of the integer.* Of course, the divisors d of n are precisely those integers of the form

$$d = p_1^{t_1} \cdots p_r^{t_r}, \quad 0 \leq t_i \leq k_i, \quad i = 1, \ldots, r.$$

Here, it is worth noting that the right-hand-side need not be quite the standard form of d since some or all of exponents may be equal to zero.

Many of the results in Chapter 2 can be proved easily by both the canonical form and the following properties.

Theorem 3.2.2. *Let $a = p_1^{k_1} \cdots p_r^{k_r} \in \mathbf{N}$ and $b = p_1^{l_1} \cdots p_r^{l_r} \in \mathbf{N}$ be the canonical form, where $k_i, l_i \in \{0\} \cup \mathbf{N}$. Then*

(i) *$a|b$ if and only if $k_i \leq l_i$ for all i, $1 \leq i \leq r$.*
(ii) $(a, b) = p_1^{\min\{k_1,l_1\}} \cdots p_r^{\min\{k_r,l_r\}}$.
(iii) $[a, b] = p_1^{\max\{k_1,l_1\}} \cdots p_r^{\max\{k_r,l_r\}}$.
(iv) *$(a, b) = 1$ if and only if $k_i l_i = 0$ for all i, $1 \leq i \leq r$.*

PROOF. We just prove that (iv) holds, and leave others as exercises. If $(a, b) = 1$, then (ii) gives

$$p_1^{\min\{k_1,l_1\}} \cdots p_r^{\min\{k_r,l_r\}} = 1$$

and so $\min\{k_i, l_i\} = 0$ for all i. Namely, $k_i = 0$ or $l_i = 0$. Conversely, if $k_i l_i = 0$, then (ii) implies $(a, b) = 1$. □

Corollary 3.2.1. *Let $a, b, c \in \mathbf{N}$. If $a|c$, $b|c$ and $(a, b) = 1$, then $ab|c$.*

PROOF. Let a and b be of the form in Theorem 3.2.1 and let

$$c = p_1^{m_1} \cdots p_r^{m_r} \quad \text{where} \quad m_i \in \{0\} \cup \mathbf{N} \quad \text{for} \quad 1 \leq i \leq r.$$

Then

$$ab = \prod_{i=1}^{r} p_i^{k_i + l_i}.$$

Since $a|c$ and $b|c$, $k_i, l_i \leq m_i$. In order to prove $ab|c$, by Theorem 3.2.2 (i) it suffices to check that $k_i + l_i \leq m_i$. Note that $(a, b) = 1$. So it follows from Theorem 3.2.2 (iv) that $k_i l_i = 0$. So if $k_i = 0$ then the desired inequality is true. Otherwise, if $k_i \neq 0$ then $l_i = 0$ and hence one has $k_i + l_i \leq m_i$ as well. □

Next we discuss briefly the distribution of primes.

Theorem 3.2.3. *If n is a composite integer, then n has a prime factor not greater than $\sqrt{n}$.*

PROOF. n may be decomposed into $n = ab$, where $a, b \in \mathbf{N}$ and $1 < a, b < n$. Suppose $a \leq b$. So $n \geq a^2$, i.e., $a \leq \sqrt{n}$. Hence any prime factor of a will be less than or equal to $\sqrt{n}$. □

Theorem 3.2.3 may be used to find all the primes less than or equal to a given positive integer. *This procedure is called the sieve of Eratosthenes.*

Theorem 3.2.4. *For any $n \in \mathbf{N}$, there exist n consecutive composite integers.*

PROOF. Consider

$$(n+1)! + 2,\ (n+1)! + 3,\ \ldots,\ (n+1)! + n + 1.$$

Each of these is composite since

$$j|(n+1)! + j \quad \text{as} \quad 2 \leq j \leq n+1.$$

□

Recall that the harmonic series $\sum_{n=1}^{\infty} n^{-1}$ is divergent. If we delete the terms n^{-1} whenever n is composite, the resulting series still diverges.

Theorem 3.2.5. *If $\mathbf{P} = \{p_i\}_{i=1}^{\infty}$ stands for the set of all primes, then the series*

$$\sum_{p \in \mathbf{P}} \frac{1}{p} = \sum_{i=1}^{\infty} \frac{1}{p_i} = \frac{1}{2} + \frac{1}{3} + \frac{1}{5} + \cdots$$

diverges.

PROOF. First, given $j, k \in \mathbf{N}$ and the ith prime p_i, let $N_{j,k}$ be the number of positive integers $n \leq k$ which are not divisible by $p_i > p_j$ for any $i > j$. If n is such an integer, then it follows from Theorem 3.2.1, where the exponents of primes are either even $2k_i$ or odd $2l_i + 1$, that we can write $n = n_1^2 m$ where m is square free – that is – m is not divisible by the square of any prime, and hence m must be of the form

$$p_1^{b_1} p_2^{b_2} \cdots p_j^{b_j}$$

with every b_i either 0 or 1. Note that there are 2^j possible choices for m. Since $n_1 \leq \sqrt{n} \leq \sqrt{k}$, we can conclude $N_{j,k} \leq 2^j \sqrt{k}$.

Next, let $S_j = \sum_{i=1}^{j} p_i^{-1}$ be the jth partial sum of the series. If the series converges to a finite number S, we can choose j so that the remainder after j terms $S - S_j$ is less than $1/2$, namely,

$$\sum_{i=j+1}^{\infty} \frac{1}{p_i} = \frac{1}{p_{j+1}} + \frac{1}{p_{j+2}} + \cdots < \frac{1}{2}.$$

Note that the number of $n \leq k$ which are divisible by each p_i is at most k/p_i. Hence $k - N_{j,k}$, the number of $n \leq k$ divisible by one or more of p_{j+1},

$p_{j+1}, \ldots,$ is not greater than

$$\frac{k}{p_{j+1}} + \frac{k}{p_{j+1}} + \cdots = k \sum_{i=j+1}^{\infty} \frac{1}{p_i} < \frac{k}{2}.$$

Hence $k - N_{j,k} < k/2$, that is, $N_{j,k} > k/2$ and consequently,

$$\frac{k}{2} < N_{j,k} \leq 2^j \sqrt{k}.$$

Clearly, the inequality $k/2 \leq 2^j \sqrt{k}$ is false for $k \geq 2^{2j+2}$. This proves the assertion. □

3.3. The Bracket Function

Let us introduce the following definition.

Definition 3.3.1. For $x \in \mathbf{R}$, let $[x]$ be the integral part of x, i.e., the largest integer which does not exceed x. This function is often called the bracket function.

For example, $[5/3] = 1$, $[\sqrt{5}] = 2$, $[-\pi] = -4$ and $[8] = 8$.

Remark 3.3.1. Below are a few words about the basic properties of the bracket functions.

(i) It is clear that $x = [x] + \alpha$, where $0 \leq \alpha < 1$.

(ii) If $a \in \mathbf{Z}$ with $a \leq \alpha \in \mathbf{R}$, then $a \leq [\alpha]$. Indeed, in case $a + 1 > \alpha$, one has $a + 1 > \alpha \geq a$ and then $a = [\alpha]$; in case $a + 1 \leq \alpha$, one gets $a \leq \alpha - 1 < [\alpha]$. Consequently, if $\alpha \leq \beta$ with $\alpha, \beta \in \mathbf{R}$, then $[\alpha] \leq [\beta]$ due to $[\alpha] \leq \alpha \leq \beta$.

(iii) Of course, $[\alpha + \beta] \geq [\alpha] + [\beta]$.

(iv) Finally, if $a \in \mathbf{Z}$, $b \in \mathbf{N}$ and $a = qb + r$, $0 \leq r < b$, then $a/b = q + r/b$, $0 \leq r/b < 1$, and it follows that $[a/b] = q$. That is, $[a/b]$ is the quotient in the division of a and b. An equivalent statement is that $[a/b]$ is the largest integer k such that $bk \leq a$.

For later use, we state the following further fact on the bracket function.

Lemma 3.3.1. *If $x \in \mathbf{R}$ and $n \in \mathbf{N}$, then $[[x]/n] = [x/n]$.*

Proof. By Definition 3.3.1, we have

$$[x/n] \leq x/n < [x/n] + 1,$$

whence producing

$$n[x/n] \leq x < n[x/n] + n,$$

and consequently,

$$n[x/n] \leq [x] < n[x/n] + n, \quad \text{i.e.,} \quad [x/n] \leq [x]/n < [x/n] + 1.$$

This proves the desired formula. □

Definition 3.3.2. Given a prime p and an integer $a > 1$, denote by $K_p(a)$ the largest integer t such that $p^t | a$.

The following formula is of independent interest.

Theorem 3.3.1. *If $n \in \mathbf{N}$, then $K_p(n!) = \sum_{s=1}^{\infty} [n/p^s]$.*

PROOF. Note first that there are only a finite number of nonzero terms in the series on the right since

$$[n/p^s] = 0 \quad \text{whenever} \quad p^s > n.$$

Next, we see that the multiples of p which occur in the product defining $n!$ are precisely

$$p,\ 2p,\ 3p,\ \ldots,\ [n/p]p,$$

and hence $K_p(n!)$ is the number of times p occurs as a factor in the product of these numbers. It follows that

$$K_p(n!) = [n/p] + K_p([n/p]!).$$

In this formula, if we replace n by $[n/p]$ and use Lemma 3.3.1, then we obtain

$$K_p([n/p]!) = [n/p^2] + K_p([n/p^2]!),$$

so that

$$K_p(n!) = [n/p] + [n/p^2] + K_p([n/p^2]!).$$

A repetition of this process finally shows the desired formula. □

Example 3.3.1. $K_3(500!) = 247$.

Solution. In fact, the simple calculations show

$$\begin{aligned}
\left[\frac{500}{3}\right] &= 166,\\
\left[\frac{500}{3^2}\right] &= [166/3] = 55,\\
\left[\frac{500}{3^3}\right] &= [55/3] = 18,\\
\left[\frac{500}{3^4}\right] &= [18/3] = 6,\\
\left[\frac{500}{3^5}\right] &= [6/3] = 2.
\end{aligned}$$

Since the sum of these is 247, 3 appears exactly 247 times as a factor in 500!. We are done.

For a direct application of Theorem 3.3.1 we derive the following useful result.

Corollary 3.3.1. *If $n, r \in \mathbf{N}$ with $1 \le r < n$, then $\frac{n!}{r!(n-r)!} \in \mathbf{N}$.*

PROOF. Let us first observe that if $x, y \in \mathbf{R}$ then $[x+y] \ge [x] + [y]$. In particular, we have that for each prime p and each $s \in \mathbf{N}$,

$$[n/p^s] \ge [r/p^s] + [(n-r)/p^s].$$

It follows from Theorem 3.3.1 that

$$K_p(n!) \geq K_p(r!) + K_p((n-r)!),$$

and so that p occurs at least as many times in $n!$ as in $r!(n-r)!$. Since this is true for every prime, it follows that $r!(n-r)!$ is a divisor of $n!$, and the proof is completed. □

Note that the number in Corollary 3.3.1 is just the coefficient of $x^{n-r}y^r$ in the binomial expansion of $(x+y)^n$. So it is equal to

$$n(n-1)\cdots(n-r+1)/r!.$$

From this it follows that for each $r \in \mathbf{N}$, the product of any r consecutive positive integers is divisible by $r!$. Moreover, if p is prime and $0 < r < p$, then $p!$ is divisible by p, but neither $r!$ nor $(p-r)!$ can be divisible by p. Accordingly, in the binomial expansion of $(x+y)^p$ all coefficients except those of x^p and y^p are divisible by p.

3.4. Mersenne, Fermat, and Perfect Numbers

Various methods have been developed for determining whether certain special types of numbers are prime or composite, and the largest primes known have been discovered in this way. In what follows, we consider the numbers introduced by Marin Mersenne and Pierre de Fermat, but also discuss the perfect numbers.

Definition 3.4.1. Numbers of the form $2^n - 1$, $n \in \mathbf{N}$ are called Mersenne numbers and primes of this form are called Mersenne primes.

Theorem 3.4.1. *Let $n-1, a-1 \in \mathbf{N}$. If $a^n - 1$ is a prime, then $a = 2$ and n is a prime.*

PROOF. Since

$$a^n - 1 = (a-1)(a^{n-1} + \cdots + a + 1)$$

and the second factor is greater than 1, it follows that $a = 2$. Otherwise, $a - 1$ would also exceed 1 and $a^n - 1$ would be composite. Moreover, if n were composite so that $n = rs$ with $r, s > 1$, then

$$2^{rs} - 1 = (2^r - 1)(2^{r(s-1)} + \cdots + 2^r + 1)$$

and each factor on the right clearly exceeds 1. This is again a contradiction, so n must be a prime. □

A statement similar to that of Mersenne was made by Fermat, who conjectured that $2^{2^n} + 1$ is a prime for any nonnegative integer n. This is certainly true for $n = 0, 1, 2, 3$ and 4. But in 1732, Leonhard Euler showed that $2^{2^5} + 1$ is divisible by 641 and so is not a prime. Therefore, the following concept appears naturally.

Definition 3.4.2. Numbers of the form $2^{2^n} + 1$ are called Fermat numbers and primes of this form are called Fermat primes.

Indeed, in spite of extensive theoretical and computational efforts, no more Fermat primes have been discovered and many number theorists suspect that no more exist. Still, Fermat's guess was not unreasonable, as the following theorem shows.

Theorem 3.4.2. *Let $n, a - 1 \in \mathbf{N}$. If $a^n + 1$ is prime, then a is even and $n = 2^r$ for some $r \in \mathbf{N}$.*

PROOF. If a were odd, then $a^n + 1 \geq 4$ would be even and so would not be a prime. Moreover, suppose n had an odd factor greater than 1, say $n = mq$ with q odd and $q > 1$. Then

$$a^n + 1 = (a^m + 1)(a^{m(q-1)} - a^{m(q-2)} + \cdots - a^m + 1).$$

Since $q \geq 3$, both factors are greater than 1 and this contradicts the fact that $a^n + 1$ is a prime. Therefore, n has no odd factor and so must be of the form $n = 2^r$ for some nonnegative integer r. □

Motivated by Theorem 3.4.2, we give the following definition.

Definition 3.4.3. A positive integer a is called perfect if $2a$ is equal to $\sigma(a)$ – the sum of the positive divisors of a.

Theorem 3.4.3. *Let $2^n - 1$ be a prime. Then $a \in \mathbf{N}$ is even perfect if and only if $a = 2^{n-1}(2^n - 1)$.*

PROOF. Let $a = 2^{n-1}(2^n - 1)$, where $2^n - 1$ is a prime. Then, a is even. By Theorem 3.2.1, the only factors of a are products of factors of $2^n - 1$ and 2^{n-1}. So, these divisors are

$$1, 2, 2^2, \ldots, 2^{n-1}, (2^n - 1), 2(2^n - 1), \ldots, 2^{n-1}(2^n - 1).$$

Consequently,

$$\sigma(a) = 2^n(2^n - 1) = 2a,$$

and it follows that a is perfect.

Conversely, suppose that a is an even perfect number. Determine n and m by $a = m2^{n-1}$ and the conditions $n \geq 2$, $m > 0$ odd. Since a is perfect, we have again

$$m2^n = 2a = \sigma(a) = \sigma(m2^{n-1}) = \sigma(m)\sigma(2^{n-1}) = \sigma(m)(2^n - 1).$$

Therefore,

$$\sigma(m) = m2^n/(2^n - 1) \in \mathbf{N}.$$

Note that $(2^n, 2^n - 1) = 1$. So it follows from Lemma 2.4.1 that $(2^n - 1)|m$ and hence that $(m/(2^n - 1))|m$. Moreover,

$$\sigma(m) = m + \frac{m}{2^n - 1},$$

so that $\sigma(m)$ is equal to the sum of m and one other positive divisor of m. Thus, m must have only two positive divisors, and so must be a prime. Also, it must be the case that $m/(2^n - 1) = 1$. Therefore, $m = 2^n - 1$ and $a = 2^{n-1}(2^n - 1)$, where $2^n - 1$ is a prime. □

Exercises

3.1. Prove that if p is an odd prime then there exist unique natural numbers x and y such that $p = x^2 - y^2$.

3.2. Prove that there are infinitely many primes of the form $4k - 1$.

3.3. Show that there are no integers x and y such that $x^2 = 2y^2$.

3.4. Write out the canonical factorizations of 4725 and 3234 and then calculate $(4725, 3234)$ and $[4725, 3234]$.

3.5. Let $n = p_1^{k_1} \cdots p_r^{k_r}$ be the canonical decomposition of a given $n \in \mathbf{N}$. Deduce a formula for the sum of the squares of the positive divisors of n. Prove that n is the square of an integer if and only if each r_j is even.

3.6. Prove that the number of positive divisors of a positive integer a is odd if and only if a is the square of the integer.

3.7. Suppose $a, b, c \in \mathbf{N}$. Prove $\frac{(a+b+c)!}{a!b!c!} \in \mathbf{N}$.

3.8. Evaluate: $[2.7 + 0.5]$; $[-75/4]$; $[-\sqrt{2} + 0.5]$.

3.9. Let p be a prime and $m, n \in \mathbf{N}$. Show

$$p|(m^p - n^p) \Rightarrow p^2|(m^p - n^p).$$

3.10. Prove the following statements:

(i) No integer is nearer n than $[n + 1/2]$.

(ii) If two integers a and b are equally near n, i.e., a and b satisfy $|a - n| = |n - b|$, then $[n + 1/2]$ is not the larger of the two integers.

3.11. Denote by $\mathcal{F}_n$ the nth Fermat number $2^{2^n}+1$. Prove $(\mathcal{F}_n, \mathcal{F}_{n+m}) = 1$ for each pair of natural numbers n and m.

3.12. Prove by exhibiting the other factor that if $a \in \mathbf{N}$ and n is even then $a + 1$ is also a factor of $a^n - 1$. Is $a + 1$ a factor of $a^n + 1$?

3.13. For a Mersenne prime $n = 2^k - 1$ let $m = n2^{k-1}$. Give all the divisors of n and show that n is a perfect number.

3.14. Prove that if

$$a, b_1, b_2, \ldots, b_n \in \mathbf{N} \quad \text{with} \quad (a, b_1) = (a, b_2) = \cdots = (a, b_n) = 1,$$

then $(a, b_1 b_2 \cdots b_n) = 1$.

3.15. For $n \in \mathbf{N}$ let $n(n+1)/2 = \sum_{j=1}^{n} j$ be the nth *triangular number.* Show that every even perfect number is a triangular number and work out all triangular numbers of the form $2^{n-1}(2^n - 1)$.

3.16. Let $a = \prod_{i=1}^{r} p_i^{k_i}$ with $k_i \in \mathbf{N}$ be the canonical factorization of $a \in \mathbf{N}$. Prove

$$\sigma(a) = \prod_{i=1}^{r} \frac{p_i^{k_i+1} - 1}{p_i - 1}.$$

3.17. Give explicitly the first seven perfect numbers and prove that no perfect number is a power of a prime.

3.18. The Prime Number Theorem says that *if* $\pi(x)$ *is the number of primes less than or equal to* $x \in \mathbf{N}$ *then*

$$\lim_{x\to\infty} \frac{\pi(x)\ln x}{x} = 1.$$

Let $li(x) = \int_2^x \frac{dt}{\ln t}$. Prove

$$\lim_{x\to\infty} \frac{\pi(x)}{li(x)} = 1.$$

Hint: Use L'Hospital's Rule to check

$$\lim_{x\to\infty} \frac{li(x)\ln x}{x} = 1.$$

CHAPTER 4

Congruences

The idea of congruences plays an important role in the theory of numbers. In this chapter, we present some of its fundamental properties such as theorems of Fermat and Euler.

4.1. Simple Properties

We start with the following notion.

Definition 4.1.1. Let $m \in \mathbf{N}$. If $a, b \in \mathbf{Z}$ are such that $m|(a-b)$, we say that a is congruent to b modulo m and write $a \equiv b \pmod{m}$. In this case, b is called a residue of a modulo m. If a is not congruent to b, that is, $m \not| (a-b)$, then we write $a \not\equiv b \pmod{m}$.

For instance, $7 \equiv 1 \pmod{3}$, $6 \equiv -2 \pmod{4}$, and $a \equiv 0 \pmod{m}$ if and only if $m|a$.

Congruences have many properties in common with equalities. Some properties are listed in the following theorem.

Theorem 4.1.1. *Let $a, b, c \in \mathbf{Z}$ and $m \in \mathbf{N}$. Then*

(i) $a \equiv b \pmod{m}$, $b \equiv a \pmod{m}$, *and* $a - b \equiv 0 \pmod{m}$ *are equivalent statements.*
(ii) *If* $a \equiv b \pmod{m}$ *and* $b \equiv c \pmod{m}$, *then* $a \equiv c \pmod{m}$.
(iii) *If* $a \equiv b \pmod{m}$ *and* $c \equiv d \pmod{m}$, *then* $a + c \equiv b + d \pmod{m}$.
(iv) *If* $a \equiv b \pmod{m}$ *and* $c \equiv d \pmod{m}$, *then* $ac \equiv bd \pmod{m}$.
(v) *If* $a \equiv b \pmod{m}$ *and* $d|m$, $d \in \mathbf{N}$, *then* $a \equiv b \pmod{d}$.
(vi) *If* $a \equiv b \pmod{m}$ *and* $c \in \mathbf{N}$, *then* $ac \equiv bc \pmod{mc}$.

PROOF. These properties follow easily from the definition. □

It is trivial that $a \equiv a \pmod{m}$ – the *reflexive* property. The properties stated in (i) and (ii) of Theorem 4.1.1 are called the *symmetric and transitive* ones, respectively; moreover (iii) and (iv) can easily be extended by induction to give the case of the finite terms, as well as the following result.

Theorem 4.1.2. *Let $a, b \in \mathbf{Z}$ and $m \in \mathbf{N}$ with $a \equiv b \pmod{m}$. Then*

(i) *If* $c \equiv d \pmod{m}$ *then* $ax + cy \equiv bx + dy \pmod{m}$ *for any* $x, y \in \mathbf{Z}$. *More generally, if* $x_1, \ldots, x_n \in \mathbf{Z}$ *and*

$$a_1 \equiv b_1 \pmod{m}, \ldots, a_n \equiv b_n \pmod{m} \quad \text{for} \quad n \in \mathbf{N},$$

then

$$\sum_{i=1}^{n} a_i x_i \equiv \sum_{i=1}^{n} a_i x_i \pmod{m}.$$

(ii) *If f is a polynomial with integer coefficients, then*

$$f(a) \equiv f(b) \pmod{m}.$$

PROOF.
(i) Assume $a - b = mk$ and $c - d = ml$. Then

$$(ax + cy) - (bx + dy) = (a - b)x + (c - d)y = m(kx + ly)$$

implies the desired result. The more general case can be verified by induction.

(ii) Let $f(x) = c_0 x^n + c_1 x^{n-1} + ... + c_n$ where $c_i \in \mathbf{Z}$. Since $a \equiv b \pmod{m}$, we can apply Theorem 4.1.1 (iv) repeatedly to find $a^k \equiv b^k \pmod{m}$ for all integers k, $1 < k \leq n$, and hence from (i) it follows that $f(a) \equiv f(b) \pmod{m}$. □

Example 4.1.1.

(i) Show $37|2^{18} + 1$ via congruences.
(ii) Find the remainder when $11|(278^5 - 3 \cdot 278^2 + 17)$.

Solution. (i) Since

$$2^6 \equiv 64 \equiv -10 \pmod{37},$$

Theorem 4.1.2 (ii) gives

$$2^{18} \equiv (2^6)^3 \equiv (-10)^3 \equiv -1000 \equiv -1 \pmod{37},$$

and consequently,

$$2^{18} + 1 \equiv -1 + 1 \equiv 0 \pmod{37}.$$

(ii) Note that $278 \equiv 3 \pmod{11}$ so by Theorem 4.1.2 (ii),

$$\begin{aligned} 278^5 - 3 \cdot 278^2 + 17 &\equiv 3^5 - 3 \cdot 3^2 + 6 \pmod{11} \\ &\equiv 3 \cdot 9^2 - 3 \cdot 9 + 6 \pmod{11} \\ &\equiv 2 \pmod{11}. \end{aligned}$$

Hence the required remainder is 2.

We are, of course, well aware of the following property of $\mathbf{R}$:

$$ax = ay \quad \text{and} \quad a \neq 0 \Rightarrow x = y.$$

Nevertheless, more care must be used in dividing a congruence through by a.

Theorem 4.1.3. *Let $a, b, c, d \in \mathbf{Z}$ and $m \in \mathbf{N}$.*

(i) *If $d > 0$, then $a \equiv b \pmod{m}$ is equivalent to $ad \equiv bd \pmod{m}$.*
(ii) *$ab \equiv ac \pmod{m}$ if and only if $b \equiv c \pmod{m/(a, m)}$.*

(iii) *If* $m_1, m_2 \in \mathbf{N}$, *then*

$$a \equiv b \pmod{m_1} \quad \textit{and} \quad a \equiv b \pmod{m_2}$$

is equivalent to $a \equiv b \pmod{[m_1, m_2]}$.

PROOF.
(i) This is a simple exercise in divisibility.
(ii) If $ab \equiv ac \pmod m$, then there is $y \in \mathbf{Z}$ such that

$$ab - ac = ym \quad \text{and hence} \quad (a/d)(b-c) = y(m/d)$$

where $d = (a, m)$. Since $(a/d, m/d) = 1$, Lemma 2.4.1 implies

$$(m/d)|(b-c), \quad \text{i.e.,} \quad b \equiv c \pmod{m/(a,m)}.$$

Conversely, suppose $b \equiv c \pmod{m/(a,m)}$; multiplying both sides by a we get $ab \equiv ac \pmod{am/(a,m)}$ by use of Theorem 4.1.1 (vi). But (a, m) is a divisor of a, so we can write $ab \equiv ac \pmod m$ by Theorem 4.1.1 (v).

(iii) If $a \equiv b \pmod{m_1}$ and $a \equiv b \pmod{m_2}$, then $m_1, m_2|(a-b)$. That is, $a - b$ is a common multiple of m_1 and m_2, and thus $[m_1, m_2]|(a-b)$. On the other hand, if $a \equiv b \pmod{[m_1, m_2]}$, then

$$a \equiv b \pmod{m_1} \quad \text{and} \quad a \equiv b \pmod{m_2}$$

follow right away due to $m_1, m_2|[m_1, m_2]$. □

In particular, Theorem 4.1.2 (ii) yields the *cancellation law*: If $(a, m) = 1$ then

$$ab \equiv ac \pmod m \Rightarrow b \equiv c \pmod m.$$

Of course, Theorem 4.1.2 (iii) can extend to the finite terms.

4.2. Residue Systems

In dealing with integers modulo m, we are essentially performing the ordinary operations of arithmetic, but we are disregarding multiples of m. In a sense we are not distinguishing between a and $a + mx$, where $x \in \mathbf{Z}$. Given $m \in \mathbf{N}$ and $a \in \mathbf{Z}$, let q and r be such that $a = mq+r$ with $0 \le r < m$; then $m|(a-r)$ and $a \equiv r \pmod m$. *In this case, r is called the least residue of a modulo m.* In fact, the forthcoming assertion holds.

Theorem 4.2.1. *Let $a, b \in \mathbf{Z}$ and $m \in \mathbf{N}$. Then $a \equiv b \pmod m$ if and only if a and b have the same least residue modulo m.*

PROOF. Since the sufficiency is obvious, it is enough to consider the necessity. Let r and r' be the least residues of a and b modulo m, respectively. Then there are q and q' such that $a = qm + r$ and $a = q'm + r'$ with $0 \le r, r' < m$. This gives $a - b = m(q - q') + r - r'$. If $a \equiv b \pmod m$, then $m|(a-b)$ and hence $m|(r - r')$. However,

$$-m < -(m-1) \le r - r' \le m - 1 < m$$

and since 0 is the only multiple of m larger than $-m$ and smaller than m, $r = r'$ follows. This proves the theorem. □

Note that Theorem 4.1.1 gives that congruences are of the reflexive, symmetric and transitive properties. So congruence of integers is an equivalence relation. For this reason, congruence divides $\mathbf{Z}$ into mutually exclusive equivalence classes or residue classes. Any integer in a given residue class is congruent modulo m to any other integer in that class, and not congruent to any integer in any other class. Furthermore, every integer is congruent to its least residue modulo m and the only least residues or remainders modulo m are the integers $0, 1, \ldots, m-1$, so it follows that every integer must be congruent modulo m to one of the numbers. Also, since no two of these remainders are congruent modulo m, there are precisely m residue classes modulo m, and the integers $0, 1, \ldots, m-1$ are representatives of these classes.

Definition 4.2.1. If $m \in \mathbf{N}$ then $\{0, 1, \ldots, m-1\}$ is called the least residue system modulo m. Any set of m integers, no two of which are congruent modulo m, is called a complete residue system modulo m.

It is clear from the above discussion that the least residue system is a complete residue system, and that a complete residue system is simply a set containing precisely one representative of each residue class modulo m.

Theorem 4.2.2. *Let $m \in \mathbf{N}$. Then a set S of integers is a complete residue system modulo m if and only if both of the following conditions are satisfied:*

(i) *S has m elements.*
(ii) *If $s, t \in S$ such that $s \equiv t \pmod{m}$, then $s = t$.*

PROOF. It is almost obvious that a complete residue system modulo m must satisfy these two conditions. Conversely, suppose that the two conditions are satisfied by S. Condition (ii) assures us that the elements of S are not congruent modulo m, that is, that no two distinct elements are congruent modulo m. Since, by condition (i), S contains m elements, exactly one element of S is in each of the m different residue classes modulo m. Since every integer is in exactly one of these residue classes, it follows that S is a complete residue system modulo m. This verifies the theorem. □

Below is an interesting remark in the algebraic point of view.

Remark 4.2.1. Given $m \in \mathbf{N}$, let $\mathbf{Z}_m$ be the set of all residue classes modulo m. If E_a denotes the residue (or congruence) class of a modulo m, that is,

$$\begin{aligned} E_a &= \{b \in \mathbf{Z} : b \equiv a \pmod{m}\} \\ &= \{\ldots, a-2m, a-m, a, a+m, a+2m, \ldots\}, \end{aligned}$$

then Theorem 4.1.1 tells us that the following addition and multiplication

$$E_a + E_b = E_{a+b} \quad \text{and} \quad E_a \cdot E_b = E_{ab}$$

are well-defined for $E_a, E_b \in \mathbf{Z}_m$. With them, it can be shown that $(\mathbf{Z}_m, +, \cdot)$ is a commutative ring, namely, $\mathbf{Z}_m$ is an Abelian group under addition, multiplication is commutative and associative, and the distributive law holds in $\mathbf{Z}_m$.

Theorem 4.2.3. *Let $b, c \in \mathbf{Z}$ and $m \in \mathbf{N}$. If $b \equiv c \pmod{m}$, then $(b, m) = (c, m)$.*

PROOF. We have $c = b + mx$ for some $x \in \mathbf{Z}$, whence establishing the desired result

$$(c, m) = (b + mx, m) = (b, m)$$

due to Lemma 2.3.1. □

Definition 4.2.2.

(i) For $a, b \in \mathbf{Z}$, we say that a is relatively prime to b provided $(a, b) = 1$. In this case, both a and b are relatively prime.

(ii) A reduced residue system modulo $m \in \mathbf{N}$ is a set of $r_j \in \mathbf{Z}$ such that $(r_j, m) = 1, r_i \not\equiv r_j \pmod{m}$ if $i \neq j$, and such that every $x \in \mathbf{Z}$ relatively prime to m is congruent modulo m to some member r_i of the set.

In view of Theorem 4.2.3 it is clear that a reduced residue system modulo m can be obtained by deleting from a complete residue system modulo m those members that are not relatively prime to m. Furthermore, all reduced residue systems modulo m will contain the same number of members, a number that is denoted by $\phi(m)$. *This function is called Euler's ϕ-function.* By applying this definition of $\phi(m)$ to the complete residue system $1, 2, \ldots, m$ mentioned above, we can get what amounts to an alternative definition $\phi(m)$. More precisely, $\phi(m)$ is the number of positive integers less than or equal to m that are relatively prime to m. In other words, we have

Theorem 4.2.4. *A set S of integers is a reduced residue system modulo $m \in \mathbf{N}$ if and only if the following three conditions are all satisfied:*

(i) *$(s, m) = 1$ for each $s \in S$.*

(ii) *S has $\phi(m)$ elements.*

(iii) *If $s, t \in S$ such that $s \equiv t \pmod{m}$, then $s = t$.*

PROOF. Our previous discussions show that a residue system must have these three properties. The converse follows easily from the following observations. Condition (i) is the initial condition for S to be a reduced residue system module m. Conditions (ii) and (iii) state that the elements of S lie in $\phi(m)$ different residue classes; hence exactly one of them must belong to each of the $\phi(m)$ different residue classes. Finally, every integer which is relatively prime to m is in one of these residue classes and hence is congruent modulo m to exactly one element of S. □

Euler's function $\phi(m)$ is of considerable interest. For instance, this function satisfies the following *product law.*

Theorem 4.2.5. *Let $m, n \in \mathbf{N}$ with $(m, n) = 1$. Then $\phi(mn) = \phi(m)\phi(n)$.*

PROOF. We put the first mn positive integers in rectangular array with m columns and n rows:

$$\begin{array}{ccccc} 1 & 2 & 3 & \cdots & m \\ m+1 & m+2 & m+3 & \cdots & m+m \\ 2m+1 & 2m+2 & 2m+3 & \cdots & 2m+m \\ \vdots & \vdots & \vdots & \vdots & \vdots \\ (n-1)m+1 & (n-1)m+2 & (n-1)m+3 & \cdots & (n-1)m+m \end{array}$$

For example, the numbers in the jth column are

$$m \cdot 0 + j,\ m \cdot 1 + j,\ m \cdot 2 + j,\ \ldots,\ m(n-1) + j.$$

By Lemma 2.3.1 we know that $(m, j) = (m, j + ma)$. Hence every element in the jth column is relatively prime to m or no element is relatively prime to m. Therefore, exactly $\phi(m)$ columns are relatively prime to m.

Consider again the jth column. Clearly there are n integers in this column and no two elements are congruent modulo n for

$$ma + j \equiv mb + j \pmod{m} \Rightarrow ma \equiv mb \pmod{n}$$

and so, by the cancellation law, $a \equiv b \pmod{n}$. Hence $a = b$ due to the fact that a and b are between 0 and $n-1$. Thus the n elements in the jth column are congruent to all the remainders $0, 1, 2, \ldots, n-1$.

Therefore, the jth column contains exactly $\phi(n)$ numbers relatively prime to n. Hence in each of the $\phi(m)$ columns containing the numbers relatively prime to m, there are $\phi(n)$ numbers relatively prime to n, and these are the only numbers relatively prime to both m and n. Here we have used the fact that $(a, mn) = 1$ if and only if $(a, m) = 1 = (a, n)$. That is, there are exactly $\phi(m)\phi(n)$ numbers in the above array that are relatively prime to both m and n. So $\phi(mn) = \phi(m)\phi(n)$. □

Remark 4.2.2. By induction it follows that if $m_1, \ldots, m_n \in \mathbf{N}$ are relatively prime in pairs then

$$\phi(m_1 \cdots m_n) = \phi(m_1) \cdots \phi(m_n).$$

In order to give a formula of ϕ, we need the following lemma.

Lemma 4.2.1. *Let p be a prime and $k \in \mathbf{N}$. Then*

$$\phi(p^k) = p^k - p^{k-1}.$$

PROOF. Recall that $\phi(p^k)$ is the number of integers x such that

$$1 \le x \le p^k \quad \text{and} \quad (x, p^k) = 1.$$

There are p^k integers between 1 and p^k, and we must count all of them except the multiples of p, namely,

$$p,\ 2p,\ 3p,\ \ldots,\ p^{k-1}p.$$

There are clearly p^{k-1} multiples of p. So

$$\phi(p^k) = p^k - p^{k-1} = p^k\left(1 - \frac{1}{p}\right).$$

□

Theorem 4.2.6. *Let $n \in \mathbf{N}$.*

(i) *If $n = 1$ then $\phi(n) = 1$.*
(ii) *If $n > 1$, then $\phi(n) = n\prod_{p|n}\left(1 - \frac{1}{p}\right)$.*

PROOF. It suffices to prove (ii). For $n > 1$ we can write $n = p_1^{k_1}\cdots p_r^{k_r}$ in canonical form. Note that $(p_i^{k_i}, p_j^{k_j}) = 1$ for $i \neq j$, so we can apply Lemma 4.2.1, Theorem 4.2.5 and Remark 4.2.2 to derive

$$\phi(n) = \prod_{j=1}^{r}\phi\left(p_j^{k_j}\right) = n\prod_{j=1}^{r}\left(1 - \frac{1}{p_j}\right),$$

thereby giving the desired formula. □

Example 4.2.1. Evaluate

$$\phi(7),\ \phi(25),\ \phi(1000),\ \phi(2909),\ \text{and}\quad \phi(15229).$$

Solution. Clearly

$$\begin{aligned}
\phi(7) &= 7 - 1 = 6;\\
\phi(25) &= \phi(5^2) = 5(5-1) = 20;\\
\phi(1000) &= 1000(1 - 1/2)(1 - 1/5) = 400;\\
\phi(2909) &= 2909 - 1 = 2908
\end{aligned}$$

since 2909 is a prime; and

$$\phi(15229) = \phi(97 \cdot 157) = \phi(97)\phi(157) = 14976$$

since 97 and 157 are primes.

4.3. Euler's and Fermat's Little Theorems

To get a deep understanding of Euler's function $\phi(m)$, in this section we present the famous Euler's Theorem and its special consequence: Fermat's Little Theorem.

We start by considering a dilation of the complete system.

Theorem 4.3.1. *For $a, m \in \mathbf{N}$ let $(a, m) = 1$. If $\{r_1, r_2, \ldots, r_n\}$ is a complete, or a reduced, residue system modulo m, then*

$$\{ar_1, ar_2, \ldots, ar_n\}$$

is a complete, or a reduced, residue system, respectively, modulo m.

Proof. If $(r_i, m) = 1$, then $(ar_i, m) = 1$ by Lemma 2.4.1. There are the same number of

$$\{ar_1, ar_2, \ldots, ar_n\} \quad \text{as of} \quad \{r_1, r_2, \ldots, r_n\}.$$

Thus we need only show that $ar_i \not\equiv ar_j \pmod{m}$ whenever $i \neq j$. However, Theorem 4.1.3 (ii) shows that

$$ar_i \equiv ar_j \pmod{m} \Rightarrow r_i \equiv r_j \pmod{m}$$

and hence $i = j$. □

The following important Euler's Theorem is a fairly easy consequence of preceding assertion.

Theorem 4.3.2. *For $a, m \in \mathbf{N}$ let $(a, m) = 1$. Then*

$$a^{\phi(m)} \equiv 1 \pmod{m}.$$

Proof. Suppose $\{r_1, r_2, \ldots, r_{\phi(m)}\}$ is a reduced residue system modulo m. Then by Theorem 4.3.1 we see that $\{ar_1, ar_2, \ldots, ar_{\phi(m)}\}$ is also a reduced residue system modulo m. Hence, corresponding to each r_i there is one and only one ar_j such that $r_i \equiv ar_j \pmod{m}$. Moreover, different r_i will have different corresponding ar_j. This means that $ar_1, \ldots, ar_{\phi(m)}$ are just the residues modulo m of $r_1, \ldots, r_{\phi(m)}$, but not necessarily in the same order. Multiplying and using Theorem 4.1.1 (iv), we obtain

$$\prod_{j=1}^{\phi(m)} (ar_j) \equiv \prod_{j=1}^{\phi(m)} r_j \pmod{m},$$

and hence

$$a^{\phi(m)} \prod_{j=1}^{\phi(m)} r_j \equiv \prod_{j=1}^{\phi(m)} r_j \pmod{m}.$$

Now $(r_j, m) = 1$, so we can use Theorem 4.1.3 (ii) to cancel the r_j and we obtain $a^{\phi(m)} \equiv 1 \pmod{m}$. □

The special case of the foregoing Euler's Theorem is the following Fermat's Little Theorem.

Theorem 4.3.3. *Let p be a prime. Then*

(i) $a^{p-1} \equiv 1 \pmod{p}$ *when* $p \nmid a$.

(ii) $a^p \equiv a \pmod{p}$ *for every* $a \in \mathbf{Z}$.

Proof.

(i) Since $1, 2, \ldots, p-1, p$ with the exception of p are relatively prime to p, $\phi(p) = p - 1$ and the assertion follows from Theorem 4.3.2.

(ii) If $p \nmid a$, then from the first part it turns out that $a^p \equiv a \pmod{p}$. If, otherwise, $p | a$, then $a \equiv 0 \pmod{p}$ and hence $a^p \equiv 0 \pmod{p}$. So in this case it is trivial that $a^p \equiv a \pmod{p}$. In addition to this argument, we use the contradiction to prove that if $a^p \not\equiv a \pmod{p}$ then $p \nmid a(a^{p-1} - 1)$ and hence $p \nmid a$ and $p \nmid (a^{p-1} - 1)$. Note that (i) tells us $p | (a^{p-1} - 1)$ under $p \nmid a$, a contradiction. Thus we are done. $\square$

Theorem 4.3.3 simply means that if we take some integer a, multiply it by itself p times and subtract a, then the result is divisible by p. The forthcoming example helps us grasp the real meaning of this assertion.

Example 4.3.1. Check the following statements:
(i) $2^{68} \equiv 6 \pmod{19}$.
(ii) $15 | a^{25} - a$ for any $a \in \mathbf{Z}$.

Solution.

(i) Note that 19 is a prime and $19 \nmid 2$. By applying Theorem 4.3.3 (i), we obtain $2^{18} \equiv 1 \pmod{19}$. Since $68 = 18 \cdot 3 + 14$, we conclude

$$2^{68} = (2^{18})^3 \cdot 2^{14} \equiv 1^3 \cdot 2^{14} \equiv 2^{14} \pmod{19}.$$

Thanks to $2^4 \equiv -3 \pmod{19}$ and $14 = 4 \cdot 3 + 2$, we derive

$$2^{14} \equiv (2^4)^3 \cdot 2^2 \equiv (-3)^3 \cdot 2^2 \equiv -27 \cdot 4 \equiv -32 \equiv 6 \pmod{19},$$

whence producing $2^{68} \equiv 6 \pmod{19}$.

(ii) Let $a \in \mathbf{N}$. Using Theorem 4.3.3 (ii) twice, we obtain

$$a^{25} = (a^5)^5 \equiv a^5 \equiv a \pmod{5}$$

and similarly, $a^3 \equiv a \pmod{3}$, which implies

$$a^{25} = (a^3)^8 a \equiv a^9 = (a^3)^3 \equiv a^3 \equiv a \pmod{3}.$$

Accordingly, $15 | a^{25} - a$ follows.

4.4. Orders and Indices

From Euler's Theorem we can see that if a is relatively prime to m, then there always exists an $s \in \mathbf{N}$, namely, $\phi(m)$, such that $a^s \equiv 1 \pmod{m}$. This leads to the following concept.

Definition 4.4.1. Given $a, m \in \mathbf{N}$, let $(a, m) = 1$. Then the smallest positive integer r such that $a^r \equiv 1 \pmod{m}$ is called the order (or exponent) of a modulo m.

It is clear that the order of a modulo m does not exceed $\phi(m)$. If $a \equiv b \pmod{m}$ then a and b have the same order modulo m. Let us emphasize that if $a^s \equiv 1 \pmod{m}$ for some $s \in \mathbf{N}$, then $(a, m) = 1$. Thus the concept of order of a modulo m is meaningful if and only if $(a, m) = 1$. In the future, when we refer to the order of a modulo m, it is to be assumed that $(a, m) = 1$, even if it is not explicitly stated.

Theorem 4.4.1. *If a has order r modulo m and $k \in \mathbf{N}$, then $a^k \equiv 1 \pmod{m}$ if and only if $r | k$.*

Proof. It is enough to verify the necessity. If $k \in \mathbf{N}$ is such that $a^k \equiv 1 \pmod{m}$, then $k \geq r$ since r is the order of a modulo m. By the division algorithm, there exists a $q \in \mathbf{N}$ and $t \in \mathbf{Z}$ such that $k = qr + t$ with $0 \leq t < r$. Thus $a^k = a^{qr}a^t$. Since $a^k \equiv 1 \pmod{m}$ and $a^{qr} \equiv 1 \pmod{m}$ by the hypothesis, it follows from $(a, m) = 1$ and the cancellation law that

$$a^t = a^{k-qr} \equiv 1 \pmod{m}.$$

Now by the definition of the order r of a modulo m, t cannot be a positive integer less than r. Hence $t = 0$ and $k = qr$, completing the proof. □

In view of Euler's Theorem, we get from Theorem 4.4.1 that for each integer a such that $(a, m) = 1$, the order of a modulo m is a divisor of $\phi(m)$. Beyond this, we have the following result.

Theorem 4.4.2. *Let $a, r \in \mathbf{N}$. If a has order r modulo m, and i, j are nonnegative integers, then $a^i \equiv a^j \pmod{m}$ if and only if $i \equiv j \pmod{r}$. In particular, the elements of the set $\{a, a^2, \ldots, a^r\}$ are incongruent modulo m.*

Proof. If $i \neq j$, we may just as well assume that the notation is so chosen that $i > j$. Since $(a, m) = 1$, we conclude that

$$a^i \equiv a^j \pmod{m} \quad \text{if and only if} \quad a^{i-j} \equiv 1 \pmod{m}.$$

By Theorem 4.4.1, the last congruence holds if and only if $r|(i - j)$, that is, if and only if $i \equiv j\,(\text{mod} r)$. The final statement of the theorem is now an immediate consequence of the fact that the elements of the set $\{1, 2, \ldots, r\}$ are incongruent modulo r. □

Since any set of $\phi(m)$ integers which are relatively prime to m and incongruent modulo m is a reduced residue system modulo m, we find that if a has order $\phi(m)$ modulo m, then $\{a, a^2, \ldots, a^{\phi(m)}\}$ is reduced residue system modulo m. *In this case, a is called a primitive root modulo m.* It will often be convenient to say that m has a primitive root. Let us now give numerical illustrations of this concept.

Example 4.4.1.

(i) If $m = 3$ then $a = 2$ is a primitive root modulo $m = 3$ since $\phi(3) = 2$ and $2^2 \equiv 1 \pmod{3}$.

(ii) Let $m = 14$. Then $\phi(14) = 6$, so that every integer which is relatively prime to 14 has order modulo 14 a divisor of 6. Let us determine the orders of the elements of the reduced residue system

$$\{1, 3, 5, -5, -3, -1\} \quad \text{modulo} \quad 14.$$

Clearly, 1 has order 1 modulo 14. Since

$$3^2 \equiv 9 \pmod{14} \quad \text{and} \quad 3^3 \equiv 27 \equiv -1 \pmod{14},$$

we see that 3 cannot have order $1, 2$, or 3; hence it must have order 6. The following table gives the orders modulo 14 of all the elements of the reduced residue system exhibited above.

Integer	1	3	5	-5	-3	-1
Order	1	6	6	3	3	2

We observe that there exists an integer which has order modulo 14 any given divisor of $\phi(14)$. In particular,

$$3^6 \equiv 1 \equiv 5^6 \pmod{14},$$

that is, 3 and 5 are primitive roots modulo 14. Moreover, the reader may illustrate the truth of the second part of Theorem 4.4.2 in this case by verifying that each of the sets

$$\{3, 3^2, 3^3, 3^4, 3^5, 3^6\} \quad \text{and} \quad \{5, 5^2, 5^3, 5^4, 5^5, 5^6\}$$

is a reduced residue system modulo 14.

(iii) However, $m = 15$ has no primitive roots. In fact, by using $\phi(15) = \phi(3)\phi(5) = 8$ and the following table which determines the orders modulo 15 of the elements of the reduced residue system $\{1, 2, 4, 7, 8, 11, 13, 14\}$:

Integer	1	2	4	7	8	11	13	14
Order	1	4	2	4	4	2	4	2

we find that there is no integer of order $\phi(15)$ modulo 15.

We now introduce a remarkably useful tool that plays a role in number theory similar to that played by the logarithm in elementary algebra.

Definition 4.4.2. Let p be an odd prime and q a primitive root modulo p. We say that r is an index of n to the base q modulo p and write $r = ind_q n$ modulo p provided

$$n \equiv q^r \pmod{p} \quad \text{and} \quad 0 \leq r < p - 1.$$

Again, observe that $\{q, q^2, \dots, q^{\phi(p)}\}$ is a reduced residue system. So each element is relatively prime to p. When talking about q^s, $0 < p - 1 = \phi(p)$, we naturally ask $(q^s, p) = 1$.

Example 4.4.2. $ind_2 10 = 5$ since $2^5 \equiv 10 \pmod{11}$, $ind_2 3 = 8$ since $2^8 \equiv 3 \pmod{11}$, and so on. Indeed, a table of indices to base 2 modulo 11 is as follows:

n	1	2	3	4	5	6	7	8	9	10
ind_2 n	0	1	8	2	4	9	7	3	6	5

The following theorem shows how closely the theory of indices parallels the theory of logarithms.

Theorem 4.4.3. *Let p be an odd prime and q a primitive root modulo p. Also let $m \in \mathbf{N}$ with $(m, p) = (n, p) = 1$. Then*

(i) $m \equiv n \pmod{p}$ *if and only if* $\text{ind}_q m \equiv \text{ind}_q n \pmod{p-1}$.
(ii) $q^{\text{ind}_q n} \equiv n \pmod{p}$ *and* $\text{ind}_q q^r \equiv r \pmod{p-1}$.
(iii) $\text{ind}_q 1 = 0$ *and* $\text{ind}_q q = 1$.
(iv) $\text{ind}_q(mn) \equiv \text{ind}_q m + \text{ind}_q n \pmod{p-1}$.
(v) $\text{ind}_q(n^t) \equiv t\text{ind}_q n \pmod{p-1}$.

PROOF.

(i) Let $r = \text{ind}_q m$ and $s = \text{ind}_q n$. Then $0 \le r, s < p-1$. Since $(q^{p-1}, p) = 1$, it is necessary to require

$$(q^s, p) = 1 \quad \text{and} \quad (q^r, p) = 1, \quad \text{i.e.,} \quad (m, p) = 1 = (n, p).$$

Of course, this condition follows by the assumption that q is a primitive root modulo p. Then

$$q^r \equiv m \equiv n \equiv q^s \pmod{p}$$

if and only if

$$\text{ind}_q m = r \equiv s = \text{ind}_q n \pmod{p-1}$$

by Theorem 4.4.2 since q is a primitive root modulo p.

(ii) Let $s = \text{ind}_q n$ be as in part (i). Then

$$q^{\text{ind}_q n} = q^s \equiv n \pmod{p}.$$

Also let $r = \text{ind}_q m$. Then $q^r \equiv m \pmod{p}$ due to (i) and

$$\text{ind}_q q^r \equiv \text{ind}_q m \equiv r \pmod{p-1}$$

by part (i).

(iii) Since $1 \equiv q^0 \pmod{p}$, it follows that $\text{ind}_q 1 = 0$. Similarly,

$$q \equiv q^1 \pmod{p} \Rightarrow \text{ind}_q q = 1.$$

(iv) Let $r = \text{ind}_q m$ and $s = \text{ind}_q n$ be as above. Then

$$q^r \equiv m \pmod{p} \quad \text{and} \quad q^s \equiv n \pmod{p}.$$

But then

$$q^{r+s} = q^r q^s \equiv mn \pmod{p}$$

and it follows from (i) that

$$\text{ind}_q mn \equiv r + s \equiv \text{ind}_q m + \text{ind}_q n \pmod{p-1}.$$

(v) Again, if $s = \text{ind}_q n$, then

$$q^s \equiv n \pmod{p} \quad \text{and} \quad q^{st} \equiv n^t \pmod{p}.$$

However by (i) and (ii), we have

$$\text{ind}_q n^t \equiv rs \equiv t\,\text{ind}_q n \pmod{p-1}.$$

□

Example 4.4.3. Use the table of indices modulo 11 as above to solve the congruence $6x \equiv 5 \pmod{11}$.

Solution. For simplicity, we use "iff" as an abbreviation for the phrase "if and only if". So

$$\begin{aligned} 6x \equiv 5 \pmod{11} \quad &\text{iff} \quad \text{ind}_2 6x \equiv \text{ind}_2 5 \pmod{10} \\ &\text{iff} \quad 9 + \text{ind}_2 x \equiv 4 \pmod{10} \\ &\text{iff} \quad x \equiv 10 \pmod{11}. \end{aligned}$$

Exercises

4.1. Suppose $m, n \in \mathbf{N}$ are odd. Show that $m^2 + n^2$ cannot be a perfect square.

4.2. Find the least residue modulo 7 of 10^{515}.

4.3. Prove $223 | (2^{37} - 1)$ and $641 | (2^{2^5} + 1)$.

4.4. Prove that $a^5 \equiv a \pmod 3$ and $a^5 \equiv a \pmod 5$ for any $a \in \mathbf{Z}$.

4.5. Let $f(x) = 143x^3 + 27x^2 - 4x + 11$. List the least nonnegative residue of

(i) $f(2) \pmod 3$;
(ii) $f(-2) \pmod 5$;
(iii) $f(4) \pmod{12}$;
(iv) $f(17) \pmod{11}$;
(v) $f(1000) \pmod{17}$.

4.6. Show that the numbers $-30, -19, -15, -4, 15$, and 39 form a reduced residue system modulo 7.

4.7. Let $m, n \in \mathbf{N}$ satisfy $(m, n) = (m, n-1) = 1$. Prove

$$\sum_{j=0}^{\phi(m)-1} n^j \equiv 0 \pmod m.$$

4.8. If $a \in \mathbf{N}$ and p is a prime obeying $p \nmid a$, then $a^{p-1} \equiv 1 \pmod p$, by Fermat's theorem. If $p-1$ is the smallest positive value of k such that $a^k \equiv 1 \pmod p$, prove that $\{a, a^2, \ldots, a^{p-1}\}$ forms a reduced residue system modulo p.

4.9. Suppose $n - 1 \in \mathbf{N}$. Show that $\phi(n)$ is even and that the sum of the integers in any reduced system modulo n is divisible by n.

4.10. Let $a \in \mathbf{N}$ and let p be an odd prime. Prove

$$a^p \equiv b^p \pmod p \Rightarrow a^p \equiv b^p \pmod{p^2}.$$

4.11. Show that if $a, b, n \in \mathbf{N}$ satisfy $ab \equiv 1 \pmod n$ then a and b have the same order modulo n.

4.12. Determine the orders of the reduced residue system

$$\{1, 3, 5, -5, -3, -1\} \quad \text{modulo} \quad 14.$$

4.13. By direct calculations, prove that 3 is a primitive root modulo 17, and compute $\mathrm{ind}_3 28$ via establishing a table of indices to the base 3 modulo 17.

4.14. Let $a, k, m, r \in \mathbf{N}$. Prove that if the order of a modulo m is r then a^k has order $r/(k, r)$.

CHAPTER 5

Congruent Equations

As is well known, *if* $f(a) = 0$, *then* a *is said to be a solution of the equation* $f(x) = 0$. *Similarly, if* $f(a) \equiv 0 \pmod{m}$, *then* a *is said to be a solution of the congruent equation* $f(x) \equiv 0 \pmod{m}$. Based on congruences, this chapter focuses on considering the problem of finding integral solutions of the above congruence where f is a polynomial with integral coefficients.

5.1. Linear Congruences and CRT

This section is devoted to a discussion of the linear congruence equations and the corresponding CRT – Chinese Remainder Theorem.

As with equations, $f(x) \equiv 0 \pmod{m}$ *is called linear when* f *is a polynomial of first degree.* Conditions for the solvability of such congruent equations are easily given.

Theorem 5.1.1. *Let* $a, b \in \mathbf{Z}$ *and* $m \in \mathbf{N}$. *Then* $ax \equiv b \pmod{m}$ *is solvable if and only if* $(a, m)|b$. *Furthermore, if there are any solutions, then there are precisely* (a, m) *incongruent solutions.*

PROOF. Suppose $(a, m)|b$, then $b = k(a, m)$ for some $k \in \mathbf{Z}$. By Theorem 2.3.1 there are

$$r, s \in \mathbf{Z} \quad \text{such that} \quad (a, m) = ar + ms.$$

Multiplying by k, we obtain $b = akr + mks$, whence producing $akr \equiv b \pmod{m}$. Accordingly, kr is a solution of the given congruence.

Conversely, if x_0 is a solution of $ax \equiv b \pmod{m}$, then there exists $q \in \mathbf{Z}$ such that $ax_0 - b = qm$. Since $(a, m)|a, m$, this implies that $d|b$, as claimed.

Finally, suppose x_0 is a solution of the congruent equation. Since $(a, m)|a$, we conclude that

$$a(x + km/(a, m)) = ax_0 + km \cdot a/(a, m) \equiv ax_0 \equiv b \pmod{m},$$

and so that $x_0 + km/(a, m)$ is a solution of the equation for each $k \in \mathbf{Z}$. Moreover, if x_0, x_1 are both solutions of the congruent equation, then

$$ax_1 \equiv b \equiv ax_0 \pmod{m}.$$

Thus, by Theorem 4.1.3 (ii), $x_1 \equiv x_0 \pmod{m/(a, m)}$ and there must exist $k \in \mathbf{Z}$ such that $x_1 = x_0 + km/(a, m)$.

In view of the preceding paragraph, $\{x_0 + km/(a, m) : k \in \mathbf{Z}\}$ contains all solutions of the congruent equation. How many are incongruent modulo m? It is clear that the (a, m) solutions are incongruent, since no two of these can differ by a multiple of m. Moreover, for any k, $x_0 + km/(a,m)$ is congruent to one of these (a,m) solutions modulo m. To see this, note that for any k there are

$$q, r \in \mathbf{Z} \quad \text{with} \quad 0 \leq r < (a,m) \quad \text{such that} \quad k = a(a,m) + r.$$

Therefore

$$\begin{aligned} x_0 + km/(a,m) &= x_0 + qm + rm/(a,m) \\ &\equiv x_0 + rm/(a,m) \pmod{m} \end{aligned}$$

and $x_0 + rm/(a,m)$ is one of the foregoing list of (a,m) solutions. □

Example 5.1.1. Solve $x \in \mathbf{Z}$ for $51x \equiv 21 \pmod{36}$.

Solution. Since $51 \equiv 15 \pmod{36}$, the given equation can be replaced by $15x \equiv 21 \pmod{36}$. Since $(15, 36) = 3$ and $3|21$, it follows from Theorem 5.1.1 that the equation has three incongruent solutions. Indeed, since -1 is obviously a solution and $36/3 = 12$, it follows from Theorem 5.1.1 again that any number which differs from -1 by a multiple of 12 is also a solution. Thus, the three solutions in the least residue system $0, 1, \ldots, 35$ are $11, 23$ and 35.

If it is not possible to guess a solution to a linear congruence, we can use the Euclidean Algorithm and the construction of the proof of Theorem 5.1.1 to obtain the desired result. Regarding Example 5.1.1, we obtain

$$3 = 5 \cdot 15 - 2 \cdot 36, \quad \text{so that} \quad 21 = 35 \cdot 15 - 14 \cdot 36.$$

Converting this to a congruence, we obtain $35 \cdot 15 \equiv 21 \pmod{36}$, so that 35 is a solution. The other solutions are now found as before.

Another possibility to find the solution is to reduce

$$15x \equiv 21 \pmod{36} \quad \text{to} \quad 5x \equiv 7 \pmod{12}$$

by Theorem 4.1.3 (ii). With this, it is not difficult to find a number whose product with 5 is congruent to 1 modulo 12. In this case, 5 is the desired number and multiplication of both sides of the congruence by 5 yields $25x \equiv 35 \pmod{12}$. But

$$25 \equiv 1 \pmod{12} \quad \text{and} \quad 35 \equiv 11 \pmod{12},$$

so that this equivalent to $x \equiv 11 \pmod{12}$. Since this argument is reversible, $x = 11$ is a solution of the given congruence and other solutions may be found from this one, as before.

Let us take a look at one more example.

Example 5.1.2. Solve all possible $x \in \mathbf{Z}$ for the following two equations:

(i) $x^{25} \equiv 3 \pmod{11}$;
(ii) $x^{24} \equiv 3 \pmod{11}$.

Solution. (i) Note that 2 is a primitive root modulo 11. By use of indices,

$$\begin{aligned} x^{25} \equiv 3 \pmod{11} \quad &\text{iff} \quad \mathrm{ind}_2 x^{25} \equiv \mathrm{ind}_2 3 \pmod{10} \\ &\text{iff} \quad 25\mathrm{ind}_2 x \equiv 8 \pmod{10}. \end{aligned}$$

Since $(25, 10) \nmid 8$, this congruence is not solvable.

(ii) Note that 2 is a primitive root modulo 11. By use of indices,

$$\begin{aligned} x^{24} \equiv 3 \pmod{11} \quad &\text{iff} \quad \mathrm{ind}_2 x^{24} \equiv \mathrm{ind}_2 3 \pmod{10} \\ &\text{iff} \quad 24\mathrm{ind}_2 x \equiv 8 \pmod{10}. \end{aligned}$$

Since $(24, 10) = 2$ and $2|8$, this congruence has two distinct solutions modulo 10. We have

$$\begin{aligned} 24\mathrm{ind}_2 x \equiv 8 \pmod{10} \quad &\text{iff} \quad 4\mathrm{ind}_2 x \equiv 8 \pmod{10} \\ &\text{iff} \quad \mathrm{ind}_2 x \equiv 2 \pmod{5}. \end{aligned}$$

Thus, the two solutions are

$$\mathrm{ind}_2 x \equiv 2 \pmod{10} \quad \text{and} \quad \mathrm{ind}_2 x \equiv 7 \pmod{10},$$

so $x \equiv 4$ or $7 \pmod{11}$.

Example 5.1.2 has actually showed how the theory of indices can be used to obtain criteria for the existence of solutions of certain polynomial congruences of a particularly simple type. To see this, let us first prove the following lemma.

Lemma 5.1.1. *Let $a \in \mathbf{Z}$ and p be an odd prime with $(a, p) = 1$, and let q be a primitive root modulo p. Then for $n - 1 \in \mathbf{N}$, $x^n \equiv a \pmod{p}$ has a solution if and only if $(n, p-1)|\mathrm{ind}_q a$. If the congruence has a solution, then it has exactly $(n, p-1)$ incongruent solutions.*

PROOF. Taking indices, we find that

$$x^n \equiv a \pmod{p} \quad \text{is equivalent to} \quad n\mathrm{ind}_q x \equiv \mathrm{ind}_q a \pmod{p-1}.$$

Now, by Theorem 5.1.1, this linear congruence has a solution when and only when $(n, p-1)|\mathrm{ind}_q a$. Moreover, if it has a solution, then it has exactly $(n, p-1)$ incongruent solutions. Hence, in this case, there are $(n, p-1)$ different values for $\mathrm{ind}_q x$ and hence $(n, p-1)$ values of x which satisfy the congruent equation $x^n \equiv a \pmod{p}$ and are incongruent modulo p. Therefore, this congruence has $(n, p-1)$ incongruent solutions, and the proof is completed. □

Using this lemma, it is now easy to prove the Generalized Euler's Criterion as follows.

Theorem 5.1.2. *Let $a \in \mathbf{Z}$ and p be an odd prime with $(p, a) = 1$. Then $x^n \equiv a \pmod{p}$ is solvable if and only if*

$$a^{(p-1)/(n,p-1)} \equiv 1 \pmod{p}.$$

If it has a solution, then it has exactly $(n, p-1)$ incongruent solutions modulo p.

Proof. Let us put $(n, p-1) = d$. Then $p-1 = \phi_1 d$ and $(p-1)/(n, p-1) = \phi_1$ for some $\phi_1 \in \mathbf{Z}$. Suppose q is a primitive root modulo p. By use of indices, we see that

$$a^{\phi_1} \equiv 1 \pmod p \quad \text{iff} \quad \phi_1 \mathrm{ind}_q a \equiv 0 \pmod{\phi_1 d}$$
$$\text{iff} \quad d | \mathrm{ind}_q a.$$

The assertion then follows at once from Lemma 5.1.1. □

Remark 5.1.1. The linear indeterminate equation $ax + by = r$ has integer solutions if and only if $ax \equiv r \pmod b$ is solvable in $\mathbf{Z}$ when and only when $(a, b)|r$ by Theorem 5.1.1. Of course, whenever solvable, the solution set is

$$\{x_k = x_0 + kb/(a, b), y_k = y_0 - ka/(a, b) : k \in \mathbf{Z}\},$$

where $\{x_0, y_0\}$ is any solution of the equation.

So far we have been dealing with a single linear congruent equation. Considering a system of linear congruences, we may use the following well-known CRT to find the desired solutions.

Theorem 5.1.3. *Let $m_1, m_2, \ldots, m_r \in \mathbf{N}$ with $(m_i, m_j) = 1$ for $i \neq j$. Then the system*

$$x \equiv c_1 \pmod{m_1}$$
$$x \equiv c_2 \pmod{m_2}$$
$$\cdots$$
$$x \equiv c_r \pmod{m_r}$$

has a unique solution modulo $m = m_1 \cdots m_r$.

Proof. Let $M_i = m/m_i$ for each $i = 1, \ldots, r$. Then $m_j | M_i$ for $j \neq i$. Since $(m_i, m_j) = 1$ when $i \neq j$, we have $(m_i, M_i) = 1$ by Lemma 3.2.1. It follows from Theorem 5.1.1 that the equation $M_i y \equiv 1 \pmod{m_i}$ has a solution b_i for each i, i.e., $M_i b_i \equiv 1 \pmod{m_i}$. Set

$$x_0 = \sum_{i=1}^{r} c_i M_i b_i.$$

Then $x_0 \equiv c_i \pmod{m_i}$ since

$$c_i M_i b_i \equiv c_i \pmod{m_i} \quad \text{and} \quad M_j \equiv 0 \pmod{m_i} \quad \text{for} \quad j \neq i.$$

Thus, x_0 is a solution of the system.

Now, suppose x_1 and x_0 are any two solutions of the system. Then $x_1 \equiv c_i \equiv x_0 \pmod{m_i}$ and consequently $m_i | (x_1 - x_0)$. In this case, $x_1 - x_0 = k_1 m_1$. Since $(m_i, m_j) = 1$ for $i \neq j$, it follows from Lemma 2.4.1 and Exercise 3.14 that $m_2 | k_1 m_1$ and so $m_2 | k_1$ and then $k_1 = k_2 m_2$. By this, $x_1 - x_0 = k_2 m_1 m_2$. Observe that

$$(m_3, m_1 m_2) = 1 \quad \text{due to} \quad (m_3, m_1) = 1 = (m_3, m_2).$$

So we may continue the previous process, and finally we get $m | (x_1 - x_0)$ and so $x_1 \equiv x_0 \pmod m$. Therefore, if a solution exists, then it is unique modulo m. □

The preceding proof of CRT is constructive and, therefore, provides a method for solving a system of congruences.

Example 5.1.3. Find the simultaneous solution of the system:

$$\begin{aligned} x &\equiv 1 \pmod 2 \\ x &\equiv 2 \pmod 3 \\ x &\equiv 1 \pmod 5 \\ x &\equiv 5 \pmod 7 \end{aligned}$$

Solution. Here we have

$$\begin{array}{lll} m_1 = 2, & M_1 = 105, & c_1 = 1 \\ m_2 = 3, & M_2 = 70, & c_2 = 2 \\ m_3 = 5, & M_3 = 42, & c_3 = 1 \\ m_4 = 7, & M_4 = 30, & c_4 = 5 \end{array}$$

and we must solve

$$M_i y \equiv 1 \pmod{m_i} \quad \text{for} \quad i = 1, 2, 3, 4.$$

In so doing, we find

$$\begin{array}{rclcccc} 105y & \equiv & 1 \pmod 2 & \text{reduces to} & y \equiv 1 \pmod 2 & \text{so} & b_1 = 1, \\ 70y & \equiv & 1 \pmod 3 & \text{reduces to} & y \equiv 1 \pmod 3 & \text{so} & b_2 = 1, \\ 42y & \equiv & 1 \pmod 5 & \text{reduces to} & 2y \equiv 1 \pmod 5 & \text{so} & b_3 = 3, \\ 30y & \equiv & 1 \pmod 7 & \text{reduces to} & 2y \equiv 1 \pmod 7 & \text{so} & b_4 = 4. \end{array}$$

Thus, the desired solution is

$$x_0 = \sum_{i=1}^{4} c_i b_i M_i = 971.$$

This solution is unique modulo $210 = 2 \cdot 3 \cdot 5 \cdot 7$. So, any other solution differs from 971 by some multiple of 210. Conversely, any number that differs from 971 by a multiple of 210 is a solution. For example, the least positive solution is $131 = 971 - 4 \cdot 210$.

There is also a faster iterative method for solving the system. It is clear that $x = 1$ is a solution of the first and third congruences of the system. Also, since the solution of these two congruences is unique modulo 10 by CRT, any other solution must be of the form $1 + 10y$ for some $y \in \mathbf{Z}$. Now, a number of this form will also be a solution of

$$x \equiv 2 \pmod 3 \quad \text{provided} \quad 1 + 10y \equiv 2 \pmod 3.$$

Solving this for y, we obtain $y \equiv 1 \pmod 3$. Thus,

$$1 + 10y = 1 + 10 \cdot 1 = 11$$

is a solution of the first three congruences of the system. Since the solution of these congruences is unique modulo 30 thanks to CRT, $11 + 30z$ satisfies the first three congruences for any $z \in \mathbf{Z}$. Hence, to find a solution to the complete system, we seek such a z that $11 + 30z \equiv 5 \pmod 7$. We have

$$30z \equiv -6 \pmod 7 \quad \text{iff} \quad 2z \equiv -6 \pmod 7$$
$$\text{iff} \quad z \equiv -3 \equiv 4 \pmod 7.$$

Thus, $11 + 30z = 131$ is a solution of the system and any other solution is congruent to this one modulo 210.

5.2. Lagrange's and Wilson's Theorems

Unlike polynomial equations, polynomial congruences of degree n frequently have more than n incongruent solutions. Thus, the linear congruence of Example 5.1.1 has three incongruent solutions. On the other hand, if p is a prime, then the linear congruent equation $ax \equiv b \pmod p$ has one and only one solution, except in the trivial case: $p|a, b$. As the following Joseph Louis Lagrange's Theorem shows, this situation is typical of congruences with prime modulo.

Theorem 5.2.1. *If p is a prime and $f(x) = \sum_{i=1}^{n} a_i x^i$ is a polynomial of degree $n \in \mathbf{N}$ with integral coefficients and with $a_n \not\equiv 0 \pmod p$, then $f(x) \equiv 0 \pmod p$ has at most n incongruent solutions modulo p.*

PROOF. For $n = 1$, the result follows from Theorem 5.1.1.

Assume the claim of Theorem 5.2.1 is true for any polynomial of the specified type of degree $k \in \mathbf{N}$. Suppose f is of degree $k+1$ and has at least $k+2$ incongruent solutions modulo p. Let s be one of these solutions. Then $f(x) = (x - s)q(x) + r$ by the division algorithm, where $r \in \mathbf{Z}$ and $q(x)$ is a polynomial of degree k with integral coefficients. Also, the leading coefficient of $q(x)$ is $a_{k+1} \not\equiv 0 \pmod p$. By the induction assumption, $q(x)$ has at most k incongruent roots modulo p. However,

$$(s - s)q(s) + r \equiv f(s) \equiv 0 \pmod p, \quad \text{so} \quad r \equiv 0 \pmod p.$$

Therefore, for any x, we have

$$(x - s)q(x) \equiv f(x) \pmod p.$$

From this it follows that if t is any one of the incongruent solutions of $f(x) \equiv 0 \pmod p$ other than s, then $(t - s)q(t) \equiv f(t) \equiv 0 \pmod p$. Since $t \not\equiv s \pmod p$, this implies that $q(t) \equiv 0 \pmod p$ so that t is a solution of $q(x) \equiv 0 \pmod p$. However, t can be any one of the incongruent solutions of $f(x) \equiv 0 \pmod p$ other than s. Since these are at least $k+1$ in number, this contradicts the fact that $q(x) \equiv 0 \pmod p$ has at most k solutions. Therefore, the assumption that $f(x)$ has at least $k+2$ incongruent solutions modulo p must be false. □

As a corollary to Lagrange's Theorem, it is possible to deduce another important result – John Wilson's Theorem – that gives a necessary condition for p to be prime.

Theorem 5.2.2. *If p is a prime, then $(p-1)! \equiv -1 \pmod{p}$.*

PROOF. Let

$$f(x) = \prod_{i=1}^{p-1}(x-i) - (x^{p-1}-1) = c_{p-2}x^{p-2} + \cdots + c_1x + c_0.$$

In view of Fermat's Little Theorem, it is clear that $f(x) \equiv 0 \pmod{p}$ has the $p-1$ incongruent solutions $\{1, 2, \ldots, p-1\}$. But this violates Lagrange's Theorem unless

$$c_{p-2} \equiv c_{p-3} \equiv \cdots \equiv c_1 \equiv c_0 \equiv 0 \pmod{p}.$$

Therefore, for any $x \in \mathbf{Z}$, we have

$$\prod_{i=1}^{p-1}(x-i) - (x^{p-1}-1) \equiv 0 \pmod{p}$$

and then

$$\prod_{i=1}^{p-1}(x-i) \equiv x^{p-1} - 1 \pmod{p}.$$

Setting $x = p$, we obtain $(p-1)! \equiv -1 \pmod{p}$, proving the assertion. □

Remark 5.2.1. If $n > 5$ is a composite number, then $n|(n-1)!$. But this leaves the case $n = 4$ thanks to $4 \not| 3!$. To see this, suppose now $n > 5$ is composite and q is such a prime that $n = qb$. Then the numbers

$$1, \ldots, q, \ldots, 2q, \ldots, (b-1)q, \ldots, n-1$$

include $b-1$ multiples of q and hence the factorial $(n-1)!$ has at least $b-1$ factors which are multiples of q. Since $n = bq$ can be at most written as $n = q^k$ where $k = \frac{\ln n}{\ln q}$, it follows that

$$\frac{\ln n}{\ln q} = k \le b - 1 = \frac{n}{q} - 1$$

is valid except for the case $q = 2$ and $n = 4$.

Wilson's Theorem can be used to determine those primes p for which $x^2 \equiv -1 \pmod{p}$ has a solution.

Theorem 5.2.3. *Let p be a prime.*

(i) *$x^2 \equiv 1 \pmod{p}$ if and only if $x \equiv \pm 1 \pmod{p}$.*
(ii) *$x^2 \equiv -1 \pmod{p}$ is solvable if and only if*

$$p = 2 \quad \text{or} \quad p \equiv 1 \pmod{4}.$$

PROOF.

(i) This congruence may be expressed as

$$x^2 - 1 = (x-1)(x+1) \equiv 0 \pmod{p},$$

which is to say that $p|(x-1)(x+1)$. Hence $p|(x-1)$ or $p|(x+1)$. Equivalently, $x \equiv \pm 1 \pmod{p}$. Conversely, if either one of these latter congruences holds, then $x^2 \equiv 1 \pmod{p}$.

(ii) If $p = 2$ we have the solution $x = 1$. For any odd prime p, we can write Wilson's Theorem in the form

$$\left(1 \cdot 2 \cdots j \cdots \frac{p-1}{2}\right)\left(\frac{p+1}{2} \cdots (p-j) \cdots j \cdots (p-2)(p-1)\right)$$
$$\equiv -1 \pmod{p}.$$

The product on the left has been divided into two parts, each with the same number of factors. Pairing off j in the first half with $p - j$ in the second half, and noting $j(p-j) \equiv -j^2 \pmod{p}$, we have

$$\begin{aligned}\prod_{j=1}^{(p-1)/2} j(p-j) &\equiv -1 \pmod{p} \\ &\equiv \prod_{j=1}^{(p-1)/2} (-j^2) \pmod{p} \\ &\equiv (-1)^{(p-1)/2} \left(\prod_{j=1}^{(p-1)/2} j\right)^2 \pmod{p}.\end{aligned}$$

Consequently, if $p \equiv 1 \pmod{4}$, then $x = ((p-1)/2)!$ is a solution of $x^2 \equiv -1 \pmod{p}$.

Conversely, suppose there is an $x \in \mathbf{Z}$ such that $x^2 \equiv -1 \pmod{p}$. We note that for such an x, i.e., x is a solution, we have $p \nmid x$. We raise both sides of the congruence to the power $(p-1)/2$ to derive

$$(-1)^{(p-1)/2} \equiv (x^2)^{(p-1)/2} = x^{p-1} \pmod{p}.$$

By Fermat's Little Theorem, the right side here is $\equiv 1 \pmod{p}$. The left side is ± 1. If $-1 \equiv 1 \pmod{p}$, then $p = 2$. If, otherwise, $-1 \not\equiv 1 \pmod{p}$, then we deduce that $(-1)^{(p-1)/2} = 1$. Accordingly, $(p-1)/2$ is even, that is, $p \equiv 1 \pmod{4}$. □

Furthermore, Theorem 5.2.3 provides the key piece of information needed to determine which integers can written as the sum of the squares of two integers. To be more specific, we need the following lemma.

Lemma 5.2.1. *Let p be a prime and $(a, p) = 1$. Then there exist $x, y \in$ $\mathbf{N}$ such that*

$$0 < x, y < \sqrt{p} \quad \textit{and} \quad ay \equiv \pm x \pmod{p}.$$

PROOF. Let

$$S = \{am + n : 0 \leq m, n < \sqrt{p}\}.$$

Since $am + n$ takes on $([\sqrt{p}] + 1)^2 > p$ possible values, there must be two of them which are congruent modulo p; call them $am_1 + n_1$ and $am_2 + n_2$. Then

$$am_1 + n_1 \equiv am_2 + n_2 \pmod{p}.$$

If $m_1 = m_2$ then $n_1 \equiv n_2 \pmod{p}$ so that $n_1 = n_2$. But this is not possible since the pairs (m_1, n_1) and (m_2, n_2) are different. Hence $m_1 \neq m_2$, and similarly $n_1 \neq n_2$ since $(a, p) = 1$. Suppose $m_1 > m_2$; then

$$a(m_1 - m_2) \equiv n_2 - n_1 \pmod{p}.$$

Put $y = m_1 - m_2$ and $x = |n_2 - n_1|$. Clearly then

$$0 < x, y < \sqrt{p} \quad \text{and} \quad ay \equiv \pm x \pmod{p}.$$

□

This leads to the desired representation theorem.

Theorem 5.2.4. *Every prime p of the form $4n + 1 \in \mathbf{N}$ is the sum of two squares.*

PROOF. Since $p \equiv 1 \pmod 4$, by Theorem 5.2.3 there exists $a \in \mathbf{Z}$ such that $a^2 \equiv -1 \pmod{p}$. Clearly $(a, p) = 1$. Choosing $x, y \in \mathbf{N}$ such that

$$0 < x, y < \sqrt{p} \quad \text{and} \quad ay \equiv \pm x \pmod{p},$$

we must have

$$x^2 + y^2 \equiv (ay)^2 + y^2 \equiv (a^2 + 1)y^2 \equiv 0 \pmod{p}.$$

But $0 < x^2, y^2 < p$ and hence $0 < x^2 + y^2 < 2p$. Accordingly, we derive $x^2 + y^2 = p$. □

5.3. Quadratic Residues

In order to solve the quadratic congruence

$$ax^2 + bx + c \equiv 0 \pmod{m},$$

we can consider its reduced form

$$ax^2 + bx + c \equiv 0 \pmod{p}$$

for some prime p. This equivalent operation depends on the following general result.

Theorem 5.3.1. *Suppose that the standard form of $m \in \mathbf{N}$ is given by $m = p_1^{k_1} p_2^{k_2} \cdots p_r^{k_r}$. If $x \equiv c \pmod{m}$ is a solution of the polynomial congruence $f(x) \equiv 0 \pmod{m}$, then $x \equiv c \pmod{p_i^{k_i}}$, $i = 1, 2, \ldots, r$, is a solution of $f(x) \equiv 0 \pmod{p_i^{k_i}}$. Conversely, if $x \equiv c_i \pmod{p_i^{k_i}}$, $i = 1, 2, \ldots, r$, is a solution of $f(x) \equiv 0 \pmod{p_i^{k_i}}$, then there exists exactly one solution $x \equiv c \pmod{m}$ of $f(x) \equiv 0 \pmod{m}$ such that $c \equiv c_i \pmod{p_i^{k_i}}$ for $i = 1, 2, \ldots, r$.*

PROOF. Clearly, an integer is divisible by m if and only if it is divisible by all $p_i^{k_i}$. Hence $f(c) \equiv 0 \pmod{m}$ if and only if $f(x) \equiv 0 \pmod{p_i^{k_i}}$ for $i = 1, 2, \ldots, r$. The first statement of the theorem follows at once. Moreover, since the different $p_i^{k_i}$ are relatively prime in pairs, the second statement is an immediate consequence of CRT. □

Now, suppose p is odd and that $(a, p) = 1$. Since $(4, p) = 1$, it follows that $(4a, p) = 1$. Therefore, solving

$$ax^2 + bx + c \equiv 0 \pmod{p}$$

is equivalent to solving

$$4a^2x^2 + 4abx + 4ac \equiv 0 \pmod{p},$$

which can be rewritten in the form

$$(2ax + b)^2 \equiv (b^2 - 4ac) \pmod{p}.$$

This congruence can be solved when and only when we can find $x_0 \in \mathbf{Z}$ such that

$$2ax_0 + b \equiv y_0 \pmod{p},$$

where y_0 satisfies $y_0^2 \equiv b^2 - 4ac \pmod{p}$. Since $(2a, p) = 1$, the first of these congruences is always solvable. Thus, the solution depends on the solvability of a congruence of the form $y^2 \equiv k \pmod{p}$. If $k \equiv 0 \pmod{p}$, then the congruence is trivially solvable. If $k \not\equiv 0 \pmod{p}$, then it is not easy to settle the congruence. To solve this problem, we need the following terminology.

Definition 5.3.1. Let $a \in \mathbf{Z}$ and p be an odd prime with $(a, p) = 1$. Then a is said to be a quadratic residue modulo (or residue of) p if $x^2 \equiv a \pmod{p}$ has a solution; otherwise, it is a quadratic non-residue modulo (or residue of) p.

It is apparent from the definition that the quadratic residues modulo p are, essentially, the squares modulo p. Thus, 1 and 4 are quadratic residues modulo 5; $1, 4$ and 2 are quadratic residues modulo 7; and a^2 is a quadratic residue modulo any odd prime p, provided that $(a, p) = 1$.

The forthcoming notational device, due to Adrien-Marie Legendre, is very useful in the study of quadratic residues.

Definition 5.3.2. Let $a \in \mathbf{Z}$. If p is an odd prime and $(a, p) = 1$, then the Legendre symbol (a/p) is defined by

$$(a/p) = \begin{cases} 1, & \text{if } a \text{ is a quadratic residue modulo } p, \\ -1, & \text{if } a \text{ is a quadratic nonresidue modulo } p. \end{cases}$$

We may emphasize that whether or not it is explicitly mentioned, whenever we use the symbol (a/p), it is to be understood that p is an odd prime and that a is not divisible by p.

Several important properties of the Legendre symbol are collected in the next theorem.

Theorem 5.3.2. *The following statements are true:*

(i) $(a^2/p) = 1$.
(ii) *If* $a \equiv b \pmod{p}$, *then* $(a/p) = (b/p)$.
(iii) $(a/p) \equiv a^{(p-1)/2} \pmod{p}$.
(iv) $(ab/p) = (a/p)(b/p)$.
(v) $(-1/p) = (-1)^{(p-1)/2}$.

PROOF. (i) and (ii) are immediate consequences of the definition, and (iii) is a consequence of Wilson's Theorem. In fact, by Theorem 5.1.1, if $1 \le r \le p-1$, then there exists s with $1 \le s \le p-1$ such that $rs \equiv a \pmod{p}$ since $(r,p) = 1|a$.

Case 1. a is a quadratic non-residue modulo p. In this case, we have $r \neq s$ or else $r^2 \equiv a \pmod{p}$, and the set $\{1, 2, \ldots, p-1\}$ can be grouped into $(p-1)/2$ pairs r_j, s_j such that

$$r_j s_j \equiv a \pmod{p} \quad \text{for} \quad j = 1, 2, \ldots, (p-1)/2.$$

Taking the product of these congruences, we obtain, by Wilson's Theorem,

$$-1 \equiv (p-1)! = a^{(p-1)/2} \pmod{p}.$$

But $(a/p) = -1$ due to the assumption above, so

$$(a/p) \equiv a^{(p-1)/2} \pmod{p}.$$

Case 2. a is a quadratic residue modulo p. Under this condition, for some r_0 and s_0, we have $r_0 = s_0$ and $r_0^2 \equiv a \pmod{p}$. Moreover,

$$(p - r_0)^2 \equiv r_0^2 \equiv a \pmod{p}.$$

By Lagrange's Theorem, these are the only solutions to $x^2 \equiv a \pmod{p}$. Thus, excluding r_0 and $p - r_0$ from consideration, the other $p-3$ numbers in $\{1, 2, \ldots, p-1\}$ can again be grouped into $(p-3)/2$ pairs r_i, s_i for which

$$r_i s_i \equiv a \pmod{p} \quad \text{with} \quad r_i \neq s_i \quad \text{for} \quad i = 1, 2, \ldots, (p-3)/2.$$

Taking the product of these congruences and multiplying both sides of the result by $r_0(p - r_0)$, we obtain, by Wilson's Theorem again,

$$\begin{aligned} -1 &\equiv (p-1)! \pmod{p} \\ &\equiv r_0(p - r_0)a^{(p-3)/2} \pmod{p} \\ &\equiv -r_0^2 a^{(p-3)/2} \pmod{p} \\ &\equiv -a^{(p-1)/2} \pmod{p}. \end{aligned}$$

Accordingly, we also derive

$$(a/p) \equiv a^{(p-1)/2} \pmod{p}.$$

We use (iii) to prove (iv) as follows:

$$\begin{aligned}(ab/p) &\equiv (ab)^{(p-1)/2} \pmod p\\ &\equiv a^{(p-1)/2} \cdot b^{(p-1)/2} \pmod p\\ &\equiv (a/p)(b/p) \pmod p.\end{aligned}$$

Since p is an odd prime and each of (ab/p), (a/p), and (b/p) is ± 1, the last congruence implies that $(ab/p) = (a/p)(b/p)$, as required.

Finally, (v) follows from (iii) by setting $a = -1$. □

Remark 5.3.1. The proof of (iii) tells us that $x^2 \equiv a \pmod p$ has exactly two solutions, respectively no solution, if and only if

$$a^{(p-1)/2} \equiv 1 \pmod p, \quad \text{respectively} \quad a^{(p-1)/2} \equiv -1 \pmod p.$$

This is a general form of Theorem 5.2.3.

Meanwhile, let us give an interesting alternate formulation of (v). Every odd prime is of the form $4n+1$ or of the form $4n+3$. If $p = 4n+1$ or $p = 4n+3$, then $(p-1)/2$ is even or odd. Accordingly, we see that -1 is a quadratic residue of every odd prime of the form $4n+1$, and a quadratic non-residue of every odd prime of the form $4n+3$.

In view of (ii), it is only necessary to determine the quadratic character of each of the numbers in a reduced residue system modulo p to know the quadratic character of any integer a. Thus, when determining quadratic residues, we usually list only those in the system

$$\{-(p-1)/2, \ldots, -1, 1, \ldots, (p-1)/2\}.$$

Theorem 5.3.3. *If p is an odd prime, then there are precisely $(p-1)/2$ incongruent quadratic residues modulo p.*

PROOF. We determine the number of integers a with $(a,p) = 1$ for which $x^2 \equiv a \pmod p$ is solvable by seeing how many different values of a we can generate by taking different values of x. First, if $x^2 \equiv a \pmod p$ and $(a,p) = 1$, then $(x,p) = 1$. Therefore, since

$$x^2 \equiv y^2 \pmod p \quad \text{as} \quad x \equiv y \pmod p,$$

it is clear that we need to consider only those values of x in the reduced system

$$\{-(p-1)/2, \ldots, -1, 1, \ldots, (p-1)/2\}.$$

Moreover,

$$(p-x)^2 \equiv x^2 \pmod p,$$

so that the squares of the numbers in the two sets

$$\{-(p-1)/2, \ldots, -1\} \quad \text{and} \quad \{1, \ldots, (p-1)/2\}$$

are congruent in pairs, and we can further restrict out attention to the first of these two sets. But

$$1^2, 2^2, \ldots, (p-1)^2/4$$

are all incongruent modulo p due to the fact that $i^2 - j^2$ can not be divided by the old prime p since $1 \le i, j \le (p-1)/2$. Thus, there are precisely $(p-1)/2$ incongruent quadratic residues modulo p and any quadratic residue is congruent to one of the numbers

$$1^2, 2^2, \dots, (p-1)^2/4.$$

□

We continue to let p be an odd prime and, for convenience, also to put $h = (p-1)/2$. Moreover, let us denote by R the reduced residue system modulo p:

$$R = \{-h, \dots, -1, 1, \dots, h\}.$$

Finally, let q be an integer which is relatively prime to p, and let us define

$$T = \{q, 2q, \dots, hq\}.$$

Now the elements of T are relatively prime to p and incongruent modulo p; hence each of them is congruent modulo p to exactly one element of R. Using the notation which we have just introduced, we can state the following Carl Friedrich Gauss's Lemma.

Lemma 5.3.1. *If n is the number of elements of T which are congruent modulo p to negative elements of R, then $(q/p) = (-1)^n$.*

PROOF. Suppose $a_1, a_2, \dots, a_n$ are the negative elements of R which are congruent modulo p to elements of T, and that $b_1, b_2, \dots, b_m$ are the positive elements of R which are congruent modulo p to elements of T. Clearly, $m + n = h$. Moreover, since the product of all elements of T is $h!q^h$, we derive

$$h!q^h \equiv a_1 \cdots a_n b_1 \cdots b_m \pmod{p}.$$

We next observe that it is impossible to get $-a_i = b_j$, since $a_i + b_j = 0$ would imply that the sum of two elements of T, say,

$$c_i q \equiv a_i \pmod{p} \quad \text{and} \quad d_j q \equiv b_j \pmod{p} \quad \text{where} \quad 1 < c_i, d_j \le h,$$

is congruent to 0 modulo p, which gives the impossible congruence

$$(c_i + d_j)q \equiv 0 \pmod{p}$$

since $2 \le c_i + d_j < p$ and $(p, q) = 1$. Therefore the set

$$\{-a_1, -a_2, \dots, -a_n, b_1, b_2, \dots, b_m\}$$

consists of h distinct positive integers not greater than h, and we must have

$$\{-a_1, -a_2, \dots, -a_n, b_1, b_2, \dots, b_m\} = \{1, 2, \dots, h\},$$

and consequently,

$$(-1)^n \prod_{i=1}^{n} a_i \prod_{j=1}^{m} b_j = h!.$$

By the formula of $h!q^h$ above, we further conclude

$$(-1)^n h! q^h \equiv h! \pmod{p}.$$

Since $(h!, p) = 1$, we can cancel $h!$ from both sides of the last congruence, obtaining

$$q^h \equiv (-1)^n \pmod p.$$

Then, by Theorem 5.3.2 (iii), we see

$$(q/p) \equiv q^{(p-1)/2} \equiv (-1)^n \pmod p,$$

and then $(q/p) = (-1)^n$ since p is an odd prime and $(q/p) = \pm 1$. This completes the proof of the lemma. □

An essential use of Gauss's Lemma gives the following result about quadratic residues: *The law of quadratic reciprocity*, as one of the most elegant and significant theorems in number theory.

Theorem 5.3.4. *If p and q are distinct odd primes, then*

$$(p/q)(q/p) = (-1)^{(p-1)/2\cdot(q-1)/2}.$$

PROOF. We split the argument into three steps below.

Step 1. We check that if $M = \sum_{i=1}^{h}[iq/p]$ with $h = (p-1)/2$ then $(q/p) = (-1)^M$.

Let $iq = [iq/p]p + r_i$ for each $i = 1, \dots, h$, where r_i is the remainder satisfying $0 < r_i < p$. In the previous notation, if iq is congruent modulo p to a negative element a_i of R, then the least positive residue of iq modulo p is $p + a_i$, so that $r_i = p + a_i$. In fact,

$$iq \equiv a_i \pmod p \Rightarrow r_i \equiv a_i \pmod p \quad \text{and so} \quad p|(r_i - a_i);$$

furthermore,

$$r_i = p + a_i \quad \text{due to} \quad -(p-1)/2 \le a_i \le -1 \quad \text{and} \quad 0 < r_i < p.$$

If iq is congruent modulo p to a positive element b_i of R, then $r_i = b_i$. Hence,

$$\sum_{i=1}^{h} iq = p\sum_{i=1}^{h}[iq/p] + \sum_{i=1}^{n}(a_i + p) + \sum_{i=1}^{m} b_i.$$

For convenience, set

$$P = \sum_{i=1}^{h} i = \frac{h(h+1)}{2} = \frac{p^2-1}{8}.$$

Then the preceding equation reduces to

$$Pq = pM + pn + \sum_{i=1}^{n} a_i + \sum_{i=1}^{m} b_i.$$

However, we have known that by the definition of a_i and b_j,

$$P = -\sum_{i=1}^{n} a_i + \sum_{i=1}^{m} b_i.$$

Thus,

$$P(q-1) = (M+n)p + 2\sum_{i=1}^{n} a_i.$$

Now that we have this relation, it is easy to prove the following fact: $(q/p) = (-1)^M$. In fact,

$$q-1 \equiv 0 \pmod 2, \quad p \equiv 1 \pmod 2$$

and the last equations involving a_i imply that $M+n \equiv 0 \pmod 2$. It follows that M and n are both even or odd, and hence that $(-1)^M = (-1)^n$, giving the desired result by Gauss's Lemma.

Step 2. We check that if $N = \sum_{i=1}^{k}[ip/q]$ with $k = (q-1)/2$ then $(p/q) = (-1)^N$.

If we interchange the roles of p and q in Step 1, we find $(p/q) = (-1)^N$. Hence we have

$$(q/p)(p/q) = (-1)^{M+N}.$$

Step 3. We verify $M + N = hk$.

To the end, we proceed to present Ferdinand Gotthold Max Eisenstein's geometric argument as follows.

In a rectangular coordinate plane, consider the rectangle whose vertices are

$$(0,0),\ (p/2,0),\ (p/2,q/2),\ (0,q/2).$$

By a lattice point we mean a point, both of whose coordinates are integers. Since both p and q are odd, the only lattice points on the sides of this rectangle are those on the coordinate axes. We shall establish $M+N = hk$ by computing in two different ways the number of lattice points that are within the rectangle (not counting those on the sides).

We first observe that the number of such lattice points is exactly hk since $[p/2]h$ and $[q/2] = k$. Now let i be a positive integer not greater than h. The diagonal (between $(0,0)$ and $(p/2,q/2)$) has equation $y = qx/p$, and the line $x = i$ intersects this diagonal in the point $(i, iq/p)$. Since iq/p is not an integer, we see that among the lattice points within the rectangle none is on the diagonal. Let us count the number below the diagonal. The number of these that are on the line $x = i$ is clearly $[iq/p]$. Hence the number below the diagonal is

$$\sum_{i=1}^{h}[iq/p] = M.$$

A similar calculation will show that the number above the diagonal is N, and we conclude that the total number of lattice points within the rectangle is $M+N$. We showed above that this number is hk, and we have therefore proved $M + N = hk$. □

Remark 5.3.2. Since $(p-1)/2$ is even or odd accordingly as p is of the form $4n+1$ or $4n+3$, we can reformulate the reciprocity law as

$$(p/q) = (q/p) \quad \text{or} \quad (p/q) = -(q/p)$$

when at least one of p and q is of the form $4n+1$ or both p and q are of the form $4n+3$. This is because

$$(p/q) = (q/p)(-1)^{\frac{p-1}{2}\cdot\frac{q-1}{2}}.$$

Example 5.3.1. Determine whether 19 is a quadratic residue or non-residue of 283.

Solution. Using Theorem 5.3.4, Remark 5.3.2 and Theorem 5.3.2 (ii), we have

$$\begin{aligned}(19/283) &= (283/19)\cdot(-1)^{9\cdot 141}\\ &= -(283/19) = -(17/19) = -(2/17) = -1,\end{aligned}$$

whence deducing that 19 is a quadratic non-residue modulo 283.

Exercises

5.1. Show that every odd integer satisfies the congruence equation $x^{17} - x \equiv 0 \pmod{15\cdot 17\cdot 32}$.

5.2. Solve the following equations:

(i) $25x \equiv 4 \pmod{11}$;

(ii) $34x \equiv 60 \pmod{98}$.

5.3. Find the smallest natural number x between 2000 and 3000 such that the following three congruences hold:

$$x \equiv 4 \pmod 5,\quad x \equiv 5 \pmod 7,\quad x \equiv 6 \pmod 8.$$

5.4. Find all integral solutions of the equation $15x + 7y = 110$.

5.5. Given $a, b, c, d \in \mathbf{N}$, show that $ax + by + cz = d$ is solvable in integers if and only if $(a, b, c)|d$.

5.6. Find the least positive integer that leaves remainders of $2, 3$ and 2 when divided by $3, 5$, and 7, respectively.

5.7. *An integer is called square-free provided it is not divisible by the square of any prime.* Show that for any $n \in \mathbf{N}$ there exist n consecutive positive integers which are not square-free.

5.8. Find the solution of the system

$$3x \equiv 7 \pmod 5,\quad x \equiv 1 \pmod 4,\quad 5x \equiv 2 \pmod{11}.$$

5.9. For $m_1, m_2 \in \mathbf{N}$ let $m = [m_1, m_2]$. Prove that the system:

$$x \equiv n_1 \pmod{m_1},\quad x \equiv n_2 \pmod{m_2}$$

is solvable for $n_1, n_2 \in \mathbf{Z}$ if and only if $(m_1, m_2)|(n_1 - n_2)$.

5.10. Suppose p is a prime and $h + k = p - 1$. Prove

$$h!k! \equiv (-1)^{k+1} \pmod p.$$

Hint: Use Wilson's Theorem and

$$p - j \equiv -j \pmod p \quad \text{for} \quad j = 1, 2, \ldots, k.$$

5.11. Prove the following converse of Wilson's Theorem:

If $(n-1)! \equiv -1 \pmod{n}$ for $n \in \mathbf{N}$, then n is a prime.

5.12. If possible, solve $x \in \mathbf{Z}$ for $x^{25} \equiv 3 \pmod{11}$.

5.13. Determine whether each of the following congruent equations has a solution:

(i) $x^2 \equiv 7 \pmod{7}$;
(ii) $x^2 \equiv 2 \pmod{129}$;
(iii) $x^2 \equiv 3 \pmod{29}$;
(iv) $x^2 \equiv 150 \pmod{151}$.

5.14. Prove that if p is an odd prime then $(2/p) = (-1)^{(p^2-1)/8}$. Use this result and the law of quadratic reciprocity to determine whether -23 is a quadratic residue or non-residue of 59.

5.15. Find odd primes p for which 3 is a quadratic residue.

5.16. Find all possible integer solutions of

$$3x^2 + x + 3 \equiv 0 \pmod{17}.$$

CHAPTER 6

Three Additional Topics

In this chapter we present three relatively disconnected topics of number theory: Cryptography by the Caesar and RSA cipherments, Diophantine equations via Pythagorean triples and Fermat's Last Theorem, and Gaussian integers through complex numbers, which will be useful at various points in future research; see also [**3, 4**]. However, we shall limit ourselves to an introductory consideration; see also [**1, 2, 5, 6**].

6.1. Cryptography by Caesar and RSA

Cryptography (from the Greek kryptós, "hidden", and gráphein, "to write") is an interdisciplinary field, drawing from mathematics and computer science concerned with information security and related issues, particularly encryption. In this section, we briefly introduce two extremely special but useful technologies used in cryptography. The first is Caesar's cipherment (one of the earliest and simplest substitution ciphers), and the second is Rivest-Shamir-Adelman's encryption (one of relatively modern cryptographic techniques).

In cryptography, a Caesar cipher, named after Julius Caesar who used it to communicate with his generals during the Gallic wars, is a type of substitution cipher in which each letter in the plaintext is replaced by a letter some fixed number of positions further down the alphabet. For instance, let us simply replace each letter in the alphabet by the letter three places to its right with the stipulation that x be replaced by a, y by b, and z by c. Then the plaintext:

REINFORCEMENTS COMING

would become, in ciphertext,

UHLQIRUFHPHQWV FRPLQJ.

The decipherment is then easily carried out by replacing each letter in the ciphertext by the letter three places to its left. We now apply our knowledge of congruence arithmetic to construct the Caesar's method of encrypting messages. Number the letters 0 through 25 and let P and C represent plaintext and ciphertext letters, respectively, as shown.

A	B	C	D	E	F	G	H	I	J	K	L	M	N	O	P	Q	R	S	T	U	V	W	X	Y	Z
0	1	2	3	4	5	6	7	8	9	10	11	12	13	14	15	16	17	18	19	20	21	22	23	24	25

Then, in numerical form,

$$C \equiv P + 3 \pmod{26} \quad \text{and} \quad P \equiv C - 3 \pmod{26}$$

with the restriction $0 \le P, C < 26$. Thus, the message above would first be converted to the numerical plaintext

17 4 8 13 5 14 17 2 4 12 4 13 19 18   2 14 12 8 13 6

and then, via the foregoing transform, to the numerical ciphertext

20 7 11 16 8 17 20 5 7 15 7 16 22 21   5 17 15 11 16 9 .

To disguise word length, the message actually transmitted would be in the form of numbers grouped into blocks of length 5. Thus

20 7 11 16 8   17 20 5 7 15   7 16 22 21 5   17 15 11 16 9 .

would be sent. Using $P \equiv C - 3 \pmod{26}$, the recipient would transform this back to the numerical plaintext

17 4 8 13 5   14 17 2 4 12   4 13 19 18 2   14 12 8 13 6

and then to the alphabetic:

REINF   ORCEM   ENTSC   OMING.

This then can be easily reassembled into the original:

REINFORCEMENTS   COMING.

Of course, a Caesar cipher with shift constant k will use the enciphering transform

$$C \equiv P + k \pmod{26}, \quad 0 \le C < 26.$$

A generalized Caesar cipher would employ an enciphering transform:

$$C \equiv rP + s \pmod{26}, \quad 0 \le C < 26,$$

where we call r *the multiplier and* s *the shift constant.* The choice for s can be arbitrary but we need $(r, 26) = 1$ so that $rP + s$ runs through a complete residue system modulo 26 as P does and also so that r^{-1} exists, where r^{-1} is the solution to $rx \equiv 1 \pmod{26}$. Then the deciphering transformation is clearly

$$P \equiv r^{-1}(C - s) \pmod{26}, \quad 0 \le P < 26,$$

where we number the letters of the alphabet as before.

As a very powerful technique used for public key encryption, we next discuss the RSA algorithm, where the initialism stands for Ron Rivest, Adi Shamir and Len Adleman, who invented it in 1977. Since the RSA algorithm and its security are based on the difficulty of factoring very large numbers, deducing an RSA key requires an extraordinary amount of computer processing power and time.

Below are the key five points of RSA's algorithm:

(i) Choose two different large primes p and q randomly.

(ii) Compute $m = (p - 1)(q - 1)$ and $n = pq$.

(iii) Choose a natural number j as the public exponent such that

$$1 < j < m \quad \text{and} \quad (j, m) = 1.$$

(iv) Compute the secret exponent k such that

$$1 < k < m \quad \text{and} \quad jk \equiv 1 \pmod{m}.$$

(v) The public key is $\langle n, j\rangle$ and the private key is $\langle n, k\rangle$. The numbers p, q, m should also be kept secret. Here n, j, k are known as the modulus, the public/encryption exponent, the secret/decryption exponent, respectively.

Now, suppose Jay wishes to allow Alison to send him a private message over an insecure transmission medium. He takes the above points to generate two keys, public and private. Jay transmits the public key to Alison and keeps the private key secret. Suppose Alison wants to send a message to Jay. She transfers the message to a natural number $N < n$. So Alison gets N and knows $\langle n, j\rangle$ which Jay has announced. Then she calculates the ciphertext corresponding to N via

$$M \equiv N^j \pmod{n},$$

whence transmitting M to Jay. This procedure clearly encrypts the message. Now the follow-up procedure is for Jay to decrypt the message. When Jay receives M from Alison, since he knows her private key $\langle n, k\rangle$ he can recover N from M by

$$N \equiv M^k \pmod{n}.$$

With this, he can recover the original message thanks to

$$M^k \equiv (N^j)^k \equiv N^{jk} \pmod{n}.$$

Note that

$$jk \equiv 1 \pmod{m} \Rightarrow jk \equiv 1 \pmod{p-1} \text{ and } jk \equiv 1 \pmod{q-1}.$$

So Fermat's Little Theorem, i.e., Theorem 4.3.3 implies

$$N^{jk} \equiv N \pmod{p} \quad \text{and} \quad N^{jk} \equiv N \pmod{q}.$$

Also, because p and q are different primes, applying CRT to the last two congruences derives

$$N^{jk} \equiv N \pmod{pq} \quad \text{and so} \quad M^k \equiv N \pmod{n},$$

as desired.

Example 6.1.1. Below is a simple example of RSA encryption and decryption:

(i) Select primes $p = 61, q = 53$.
(ii) Compute $m = (p-1)(q-1) = 3120$ and $n = pq = 3233$.

(iii) Choose $j = 17$ and check

$$(j, p-1) = (17, 60) = 1 \quad \text{and} \quad (j, q-1) = (17, 52) = 1.$$

So

$$(j, m) = (17, 3120) = 1.$$

(iv) Find k such that $jk \equiv 1 \pmod{m}$, namely,

$$k = 2753 \Rightarrow 17k \equiv 1 \pmod{3120}.$$

(v) The public and private keys are $< 17, 3233 >$ and $< 2753, 3233 >$ respectively. And the encryption and decryption functions are

$$N \mapsto N^j \pmod{n} \equiv N^{17} \pmod{3233}$$

and

$$M \mapsto M^k \pmod{n} \equiv M^{2753} \pmod{3233},$$

respectively, *where N and M are called the plaintext and ciphertext, respectively.* To encrypt the plaintext value $N = 123$, we compute

$$N^{17} \pmod{3233} \equiv 123^{17} \pmod{3233} = 855,$$

and at the same time, in order to decrypt the ciphertext value $M = 855$, we use the following expansion

$$2753 = 1 + 2^6 + 2^7 + 2^9 + 2^{11} = 1 + 64 + 128 + 512 + 2048$$

to evaluate

$$\begin{aligned}
& M^{2753} \pmod{3233} \\
\equiv\ & 855^{2753} \pmod{3233} \\
\equiv\ & 855^{1+64+128+512+2048} \pmod{3233} \\
\equiv\ & 855^1 \cdot 855^{64} \cdot 855^{128} \cdot 855^{512} \cdot 855^{2048} \pmod{3233} \\
\equiv\ & 855 \cdot 916 \cdot 1709 \cdot 1160 \cdot 2197 \pmod{3233} \\
\equiv\ & 794 \cdot 1709 \cdot 1160 \cdot 2197 \pmod{3233} \\
\equiv\ & 2319 \cdot 1160 \cdot 2197 \pmod{3233} \\
\equiv\ & 184 \cdot 2197 \pmod{3233} \\
\equiv\ & 123 \pmod{3233} = 123.
\end{aligned}$$

Although this example seems no much more secure than using a simple Caesar substitution cipher, it serves to illustrate a simple example of the mechanics of RSA encryption. Furthermore, it is worth remarking that computing $N^j \pmod{n}$ is relatively easy, but calculating the inverse $M^k \pmod{n}$ is very difficult for large n's. However, if one can factor n into its prime factors p and q, the solution becomes easy again, even for large n's.

6.2. Diophantine Equations $x^n + y^n = z^n$, $n \geq 2$

We often encounter situations in which we wish to find solutions, i.e., the possible assignments of integers for the unknowns for which an equation between two polynomials with integer coefficients with any number of unknowns is satisfied. In this case we refer to the equation as a Diophantine equation, where the word Diophantine refers to Diophantus of Alexandria, who initiated a study of such equations. Due to the complexity of general Diophantine equations, we restrict our attention to the following typical but important equations: $x^n + y^n = z^n$, $n - 1 \in \mathbf{N}$.

As for the case $n = 2$, serving as a basis for the definition of distance between two points in the Euclidean plane, we recall the well-known Pythagorean Theorem or Pythagoras's Theorem – *the sum of the squares of the sides a and b of a right triangle equals the square of the hypotenuse c, namely, $a^2 + b^2 = c^2$ for which the three natural numbers a, b, c are called a Pythagorean triple.* Of course, the most famous triple is $a = 3, b = 4, c = 5$. With this, we find that there are infinitely many Pythagorean triples:

$$a = 3d, b = 4d, c = 5d, \quad d \in \mathbf{N}.$$

However, these triples are not so interesting since they have common factors. The question is: Are there infinitely many triples with no common factors (*these triples are then called primitive Pythagorean triples*)? The answer is 'yes'.

Theorem 6.2.1. *The Diophantine equation: $x^2 + y^2 = z^2$ has infinitely many primitive Pythagorean triple solutions: Every primitive triple (x, y, z) with x odd and y even can by obtained by using the formulas:*

$$x = 2st,\ y = s^2 - t^2,\ z = s^2 + t^2,$$

where positive integers $s > t \geq 1$ are chosen to be any odd integers with no common factors.

PROOF. Note that

$$x^2 = 4s^2t^2,\ y^2 = s^4 - 2s^2t^2 + t^4,\ z^2 = s^4 + 2s^2t^2 + t^4.$$

So we have $x^2 + y^2 = z^2$. Also if $(s, t) = 1$ with one of s and t even and the other odd, then x is even and y and z are both odd. Assume $(y, z) = d > 1$. Then there is a prime p such that $p|d$. Consequently, $p|y$, $p|z$, so p divides $z + y = 2t^2$ and $z - y = 2s^2$. But p must be odd since $p|z$ and z is odd. Therefore, $p|t^2$ and $p|s^2$, so $p|t$ and $p|s$. This contradicts $(s, t) = 1$, so it must be the case that $(y, z) = 1$. Similarly, it can be shown that $(x, y) = (x, z) = 1$. Therefore, the above-defined (x, y, z) forms a primitive Pythagorean triple.

Now, suppose that (x, y, z) is a primitive Pythagorean triple so that

$$x^2 + y^2 = z^2 \quad \text{and} \quad (x, y, z) = 1.$$

It is clear that

$$(x, y) = (x, z) = (y, z) = 1.$$

For example, if $(x, z) = g > 1$ then by the Fundamental Theorem of Arithmetic, there exists a prime p such that $p|d$. Since $g|x$ and $g|z$, it follows that

$$p|x^2 \text{ and } p|z^2 \text{ and hence } p|y^2 \text{ due to } y^2 = z^2 - x^2.$$

Since p is prime, $p|y$ and thus p is a common divisor in contradiction to $(x, y, z) = 1$. Similarly, we can show that $(x, y) = (y, z) = 1$.

From the preceding paragraph, it follows that x and y cannot both be even – otherwise, z would be even. It is also true that they cannot both be odd. This follows from the fact that the square of an odd integer must be of the form $4q + 1$ and the square of an even integer must be of the form $4q$, so that the square of an integer cannot be of the form $4q + 2$ or $4q + 3$. Thus, if x and y were both odd, then

$$x^2 = 4r + 1 \text{ and } y^2 = 4s + 1$$

for some r and s in $\mathbf{Z}$ and

$$z^2 = x^2 + y^2 = 4(r + s) + 2.$$

This says that z^2 is of the form $4q+2$, and this is impossible, as noted above. Hence it must be the case that one of x and y is even, the other odd.

For definiteness, let x and y be even and odd respectively. Of course, z^2 will then be of the form $4q + 1$, so z is also odd. Hence $z + y$ and $z - y$ are both even and

$$x^2 = z^2 - y^2 = (z - y)(z + y).$$

Let $z - y = 2u$ and $z + y = 2v$. Then $z = v + u$ and $y = v - u$, and it is easy to check that one of u and v is even and the other odd. Furthermore, we have $(u, v) = 1$: In fact, suppose $(u, v) = d > 1$. If u and v are both even, then $2|u$ and $2|v$, so $2|y$ and $2|z$, in contradiction to $(z, y) = 1$. Moreover, if u and v are both odd, then z and y are even, and this is also a contradiction.

Since x is even, $x/2$ is an integer and

$$\frac{x^2}{4} = \frac{z - y}{2} \cdot \frac{z + y}{2} = u \cdot v.$$

Let

$$\frac{x}{2} = \prod_{i=1}^{r} p_i^{a_i}$$

be the the canonical factorization of $x/2$. Then

$$u \cdot v = \prod_{i=1}^{r} p_i^{2a_i}.$$

Consequently,

$$u = \prod_{i=1}^{r} p_i^{b_i}, \quad v = \prod_{i=1}^{r} p_i^{c_i}, \quad b_i \geq 0,\, c_i \geq 0,\, b_i + c_i = 2a_i.$$

If b_i and c_i are both different from 0 for some i, then $p_i|u$ and $p_i|v$, in contradiction to $(u, v) = 1$. Thus, one of b_i and c_i is zero for each i. It follows that b_i and c_i are even for each i, say $b_i = 2u_i$ and $c_i = 2v_i$. Thus $u = s^2$ and $v = t^2$, where

$$s = \prod_{i=1}^{r} p_i^{u_i} \quad \text{and} \quad t = \prod_{i=1}^{r} p_i^{v_i}.$$

Also $(s,t) = 1$ and one of s and t is even and the other odd, since the same statements are valid for u and v.

Since $x > 0$ and $x = 2\sqrt{uv}$, it follows that if (x,y,z) is a primitive Pythagorean triple, then there exist integers s and t, with $(s,t) = 1$ and with one of s and t even and the other odd, such that

$$x = 2st, \quad y = t^2 - s^2, \quad z = t^2 + s^2.$$

Of course, we must also choose $t > s$ since $y > 0$. Therefore, the proof is complete. □

Since $x^2 + y^2 = z^2$ has infinitely many integer solutions, it is natural to investigate the situation where 2 is replaced by $3, 4, \ldots,$ and so forth. In 1637 Fermat showed that there are no integer solutions for exponent 4.

Theorem 6.2.2. *The Diophantine equation $x^4 + y^4 = z^4$ has no nontrivial integer solutions.*

PROOF. Assume $x^4 + y^4 = z^4$ where x, y, z have no common factors and y is even. Then Theorem 6.2.1 implies

$$y^2 = 2uv, \; x^2 = u^2 - v^2, \; z^2 = u^2 + v^2$$

where $u > v \geq 1$ with $(u,v) = 1$. Since $x^2 + v^2 = u^2$, u, x, v may be rewritten as (due to Theorem 6.2.1 once again)

$$v = 2ab, \; x = a^2 - b^2, \; u = a^2 + b^2$$

as above, where $a > b \geq 1$ with $(a,b) = 1$. Accordingly,

$$y^2 = 2uv = 4ab(a^2 + b^2).$$

Using the standard factorization, we further find that a, b, ab and $a^2 + b^2$ are squares since they have no common factors. Let $a = A^2$ and $b = B^2$. Then

$$A^4 + B^4 = a^2 + b^2$$

is a square and

$$A^4 + B^4 = a^2 + b^2 = u < u^2 + v^2 = z^2 < z^4 = x^4 + y^4.$$

This sets up an infinite descent chain of squares of integer numbers of the form $x^4 + y^4$ which is clearly impossible. □

In 1753 Euler introduced the imaginary numbers to verify that the answer to the case $n = 3$ is 'no' although his proof was not perfect.

Theorem 6.2.3. *The Diophantine equation $x^3 + y^3 = z^3$ has no nontrivial integer solutions.*

Proof. Suppose x, y, z are nonzero, pairwise relatively prime integers satisfying $x^3 + y^3 = z^3$. So it is easy to find out that only one of x, y, z is even. For instance, let z be even. Then x and y are odd. Among all the solutions with the above properties, we choose one for which $|z|$ is the smallest possible. By writing

$$x + y = 2a \quad \text{and} \quad x - y = 2b,$$

we have

$$x = a + b, \quad y = a - b, \quad \text{and} \quad z^3 = 2a(a^2 + 3b^2),$$

whence finding $a, b \neq 0$ and $(a, b) = 1$. Furthermore,

$$z^3 = x^3 + y^3 = (a + b)^3 + (a - b)^3 = 2a(a^2 + 3b^2).$$

Since z is even and $a^2 + 3b^2$ is odd, we conclude that $8|z^3$, $8|2a$, and b is odd. Consequently, we find that $(2a, a^2 + 3b^2)$ is either 1 or 3. This leads to a consideration of two cases:

$$(2a, a^2 + 3b^2) = 1 \quad \text{or} \quad (2a, a^2 + 3b^2) = 3.$$

However, we will only deal with the former, and leave the latter for the reader since its treatment is similar.

Suppose $(2a, a^2 + 3b^2) = 1$. Clearly, we have $3 \nmid a$. From $z^3 = 2a(a^2 + 3b^2)$ and the standard factorization of integers into primes it turns out that

$$2a = r^3 \quad \text{and} \quad a^2 + 3b^2 = s^3,$$

where s is odd and not a multiple of 3. Note that this s can be written as $u^2 + 3v^2$, with $u, v \in \mathbf{Z}$ and

$$a = u(u^2 - 9v^2) \qquad b = 3v(u^2 - v^2);$$

see also [**5**, pp. 27–31] for a proof. So v is odd, u is even but not zero, $3 \nmid u$ and $(u, v) = 1$. Accordingly, $2u$, $u + 3v$, $u - 3v$ are pairwise relatively prime, but also are cubes:

$$2u = \gamma^3, \quad u - 3v = \beta^3, \quad u + 3v = \alpha^3$$

thanks to

$$r^3 = 2a = 2u(u - 3v)(u + 3v).$$

Here α, β, γ are nonzero and pairwise relatively prime. This produces

$$\alpha^3 + \beta^3 = \gamma^3,$$

where γ is even and satisfies

$$|z|^3 = |2a(a^2 + 3b^2)| = |\gamma^3(u^2 - 9v^2)(a^2 + 3b^2)| \geq 3|\gamma|^3 > |\gamma|^3.$$

Of course, this is against the choice of $|z|$ as minimal. □

There have been many mathematicians working on this equation for $n \geq 5$, that is, trying to prove that one cannot find non-trivial triples analogous to Pythagorean triples but with exponents higher than 2. Although Fermat wrote, in 1637, in his copy of Claude-Gaspar Bachet's translation of the famous Arithmetica of Diophantus: "Cuius rei demonstrationem mirabilem sane detexi hanc marginis exiguitas non caperet", which means "I have discovered a truly remarkable proof of the higher order case that the margin of this page is too small to contain", no correct proof has been found until 1995. Below is the final answer – which is nowadays called Fermat's Last Theorem or Fermat's Great Theorem – one of the most famous theorems in the history of mathematics.

Theorem 6.2.4. *The Diophantine equation $x^n+y^n = z^n$ has no solution in integers all different from 0 when $n-2 \in \mathbf{N}$.*

PROOF. Using sophisticated tools from algebraic geometry, in particular elliptic curves and modular forms, Galois theory and Hecke algebras, Andrew Wiles, with help from Richard Taylor, devised a proof of Fermat's Last Theorem that was published in his work: Modular elliptic curves and Fermat's Last Theorem, Annals of Math., **141** (1995), 443-551.

A few more words are worthy of saying here. The story of the proof is almost as remarkable as the mystery of the theorem itself. Wiles spent seven years on figuring nearly all the details by himself. When he announced his proof over the course of three lectures delivered at Cambridge University on June 21-23, 1993, he amazed his audience with the number of ideas and constructions used in his proof. Unfortunately, upon closer inspection a serious error was discovered – it appeared to lead to the breakdown of his original proof. Together, Wiles and Taylor then spent about a year on trying to revive the argument. In September 1994, they were able to resurrect the proof with some different, discarded techniques that Wiles had used in his earlier attempts. □

6.3. Gaussian Integers

In the sequel, we introduce Gaussian integers, but also compare them with ordinary integers with respect to units, primes, the division algorithm and the Fundamental Theorem of Arithmetic.

A Gaussian integer is a complex number whose real and imaginary part are both integers. The Gaussian integers, with ordinary addition and multiplication of complex numbers, form an integral domain. This is a Euclidean domain which cannot be turned into an ordered ring. Formally, we have the following definition.

Definition 6.3.1. The set

$$\mathbf{G} = \mathbf{Z}[i] = \{a + bi : \quad a, b \in \mathbf{Z}, i = \sqrt{-1}\}$$

is called the set of Gaussian integers.

Remark 6.3.1. $\mathbf{G}$ is a subset of $\mathbf{C}$ – the complex field, and it is closed under the two operations of addition and multiplication in $\mathbf{C}$. Note also that $0, 1 \in \mathbf{G}$. In fact $(\mathbf{G}, +, \cdot)$ is a commutative ring, that is, $(\mathbf{G}, +)$ is an Abelian group, multiplication is commutative and associative, and the distributive law holds in $\mathbf{G}$. Since $\mathbf{Z} \subset \mathbf{G}$, we need to distinguish between ordinary integers and Gaussian integers. Accordingly, *the ordinary integers in* $\mathbf{Z}$ *will be called rational integers.*

Definition 6.3.2. Let $\alpha, \beta \in \mathbf{G}$ with $\alpha \neq 0$. We say α divides β and write $\alpha|\beta$ if there exists $\gamma \in \mathbf{G}$ such that $\beta = \gamma\alpha$. If $\alpha|\beta$ and $\beta|\alpha$ where $\beta \neq 0$, then α and β are called associates. If $\alpha|1$, then α is called a unit.

For example, $3 + 5i$ and $5 - 3i$ are associates.

Theorem 6.3.1. *Let* $\alpha, \beta \in \mathbf{G}$ *with* $\alpha, \beta \neq 0$. *Then* α *and* β *are associates if and only if* $\beta = \epsilon\alpha$ *for some unit* ϵ.

PROOF. It follows from Definition 6.3.2. □

Definition 6.3.3. Let $\alpha = a + bi \in \mathbf{G}$. Then $\bar{\alpha} = a - bi$ is called the conjugate of α, and $N(\alpha) = \alpha\bar{\alpha} = a^2 + b^2$ is called the norm of α.

Remark 6.3.2. It is a good exercise to prove:

(i) $\overline{\alpha\beta} = \bar{\alpha}\bar{\beta}$, and $N(\alpha\beta) = N(\alpha)N(\beta)$;
(ii) $N(\alpha) = 0$ iff $\alpha = 0$, and $N(\alpha) = 1$ iff α is a unit;
(iii) The set of units in $\mathbf{G}$ is $\{\pm 1, \pm i\}$.

The following is the division algorithm.

Theorem 6.3.2. *Given* $\alpha, \beta \in \mathbf{G}$ *with* $\alpha \neq 0$, *there exist* $\gamma, \delta \in \mathbf{G}$ *such that* $\beta = \gamma\alpha + \delta$ *where* $N(\delta) < N(\alpha)$.

PROOF. Note that

$$\beta/\alpha = (\beta\bar{\alpha})/(\alpha\bar{\alpha}) = A + Bi \quad \text{where} \quad A, B \in \mathbf{Q}.$$

Choose $a, b \in \mathbf{Z}$ such that

$$|A - a| \leq 1/2 \quad \text{and} \quad |B - b| \leq 1/2.$$

Let $\gamma = a + bi$ and $\delta = \beta - \gamma\alpha$. We need to show that $N(\delta) < N(\alpha)$. However, $N(\alpha) \neq 0$, plus a simple calculation, implies

$$\begin{aligned} N(\delta) &= N(\beta - \gamma\alpha) \\ &= N\big(\alpha(\beta/\alpha - \gamma)\big) \\ &= N(\alpha)N(\beta/\alpha - \gamma) \\ &= N(\alpha)N\big((A - a) + (B - b)i\big) \\ &= N(\alpha)\big((A - a)^2 + (B - b)^2\big) \\ &\leq N(\alpha)/2 \\ &< N(\alpha), \end{aligned}$$

as desired. □

Examining the division algorithm in $\mathbf{Z}$, we might ask why the norm map was needed in $\mathbf{G}$. It is simply an ingenious way of overcoming the problem of not having a natural ordering in $\mathbf{C}$. It makes no sense, in general, to say that one complex number is less than another. Whereas, $N(a+bi) = a^2+b^2$ is a real number and in $\mathbf{R}$, we do have a natural ordering.

Definition 6.3.4. For $\alpha, \beta \in \mathbf{G}$ not both 0, $g \in \mathbf{G}$ is called a greatest common divisor (gcd) of α and β provided

(i) γ is common divisor, i.e., $\gamma|\alpha, \beta$;
(ii) For λ any common divisor of α and β, $\lambda|\gamma$.

Remark 6.3.3. It is easy to show that if g is a *gcd* of α and β, then other *gcd*'s are $-g$, ig, and $-ig$.

Theorem 6.3.3. *Let $\alpha, \beta \in \mathbf{G}$ with $N(\alpha)N(\beta) \neq 0$. Then an element γ in the set $S = \{\alpha x + \beta y | x, y \in \mathbf{G}\}$ with least positive norm is a gcd of α and β.*

PROOF. Let $\gamma = \alpha x_0 + \beta y_0$ for some $x_0, y_0 \in \mathbf{G}$. We first prove that $\gamma|\alpha$ and $\gamma|\beta$. If $\gamma \not| \beta$, then it follows from the division algorithm that there exist $q, \delta \in \mathbf{G}$ such that

$$\beta = q\gamma + \delta, \quad 0 < N(\delta) < N(\gamma).$$

Hence

$$\delta = \beta - q\gamma = \beta - q(\alpha x_0 + \beta y_0) = \beta(1 - ay_0) + \alpha(-qx_0)$$

and consequently $\delta \in S$. This contradicts the fact that γ is an element with least positive norm in S. Thus $\gamma|\beta$ and similarly $\gamma|\alpha$, so γ is a common divisor of α and β.

Suppose next $\lambda|\alpha$ and $\lambda|\beta$. It is clear that $\lambda|\alpha x_0 + \beta y_0$ and thus $\lambda|\gamma$. Therefore, γ is a *gcd* of α and β. The proof is complete. □

From the argument of Theorem 6.3.3 and Remark 6.3.3 it turns out that if γ is a *gcd* of α and β then there are $x_0, y_0 \in \mathbf{G}$ such that $\gamma = \alpha x_0 + \beta y_0$.

Example 6.3.1. Find a *gcd* of $16 + 7i$ and $10 - 5i$.

Solution. Since

$$16 + 7i = (1 + i)(10 - 5i) + 1 + 2i \quad \text{and} \quad 10 - 5i = -5i(1 + 2i) + 0,$$

we conclude that

$$(6 - 5i)(1 + 2i) = 16 + 7i \quad \text{and} \quad -5i(1 + 2i) = 10 - 5i,$$

so that $1 + 2i$ is a *gcd* of $16 + 7i$ and $10 - 5i$. Consequently, the associates:

$$2 - i, -2 + i, -1 - 2i \quad \text{of} \quad 1 + 2i$$

are other *gcd*'s.

Definition 6.3.5. $\tau \in \mathbf{G}$ is called a Gaussian prime if τ is neither zero nor a unit and is divisible only by units and associates of itself. The original primes, by contrast, are referred to as rational primes.

It is worth pointing out that not all rational prime numbers are Gaussian primes, for example,

$$2 = (1 + i)(1 - i) \quad \text{and} \quad 5 = (2 + i)(2 - i).$$

Lemma 6.3.1. *Let $\alpha, \beta \in \mathbf{G}$ and τ be a prime. If $\tau|\alpha\beta$, then $\tau|\alpha$ or $\tau|\beta$.*

PROOF. If $\tau|\alpha$ then we are done. If $\tau \not|\alpha$, then 1 is a *gcd* of τ and α since τ is prime. Hence there exist $x_0, y_0 \in \mathbf{G}$ such that

$$1 = \tau x_0 + \alpha y_0 \quad \text{and hence} \quad \beta = \tau\beta x_0 + \alpha\beta y_0.$$

Since τ clearly divides the right side of the last equality, $\tau|\beta$. □

Remark 6.3.4. By induction, we can prove that if $\tau|\alpha_1\alpha_2\cdots\alpha_n$ then τ divides at least one factor α_i.

We are now ready to state and prove the Fundamental Theorem of Arithmetic for Gaussian integers.

Theorem 6.3.4. *If α is a nonzero and non-unit Gaussian integer, then α can be factorized into Gaussian primes and this factorization is unique apart from the order of the prime factors and multiplication by units.*

PROOF. We use induction on the norm of α. If $N(\alpha) = 2$, then α is prime for if $\alpha = \alpha_1\alpha_2$ then

$$2 = N(\alpha) = N(\alpha_1)N(\alpha_2)$$

and hence

$$N(\alpha_1) = 1 \quad \text{or} \quad N(\alpha_2) = 1,$$

which makes one of α_1 and α_2 a unit. If α is a prime then the statement is true. If α is not a prime, then

$$\alpha = \alpha_1\alpha_2 \quad \text{where} \quad 1 < N(\alpha_1), N(\alpha_2) < N(\alpha).$$

By the induction hypothesis both α_1 and α_2 can be factorized into primes and hence α can also be. Thus by the first form of the *PMI*, all nonzero and non-unit Gaussian integers can be factorized into primes.

Suppose now there is an α with two different factorizations. Dividing out any associate primes common to the two representations, we would have an equality of the form

$$\epsilon\tau_1\tau_2\cdots\tau_r = \epsilon'\tau_1'\tau_2'\cdots\tau_s'$$

where ϵ, ϵ' are units and τ_j, τ_j' are primes, not necessarily all distinct, but where no prime or its associate on the left occurs on the right side. But this is impossible. In fact, because $\tau_1|\tau_1'\tau_2'\cdots\tau_s'$, it follows from Lemma 6.3.1 and Remark 6.3.4 that τ_1 divides at least one of the $\tau_1', \ldots, \tau_s'$, say τ_j'. This forces τ_1 to be an associate of τ_j', which is a contradiction, and consequently proves the theorem. □

Now we address the problem of determining all the primes in $\mathbf{G}$.

Theorem 6.3.5. *If $N(\alpha)$ is a rational prime, then α is a Gaussian prime.*

PROOF. Suppose $\alpha = \alpha_1\alpha_2$. Then

$$p = N(\alpha) = N(\alpha_1)N(\alpha_2).$$

Since $N(\alpha_1)$ and $N(\alpha_2)$ are positive integers and p is a rational prime, we conclude that

$$\text{either} \quad N(\alpha_1) = 1 \quad \text{or} \quad N(\alpha_2) = 1,$$

and so that α_1 or α_2 is a unit. Thus α is a Gaussian prime. □

Remark 6.3.5. The converse of Theorem 6.3.5 is false because, for example, $N(3) = 9$ but 3 is a Gaussian prime – if $3 = \alpha\beta$ where α and β are not units, then

$$N(\alpha) = 3 \quad \text{and hence} \quad a^2 + b^2 = 3$$

for some rational integers a and b, but this is impossible.

Theorem 6.3.6.

(i) *If p is a rational prime, then p is a Gaussian prime whenever $p \equiv 3 \pmod 4$ or else there is a Gaussian prime τ such that $N(\tau) = p$ whenever $p = 2$ or $p \equiv 1 \pmod 4$.*

(ii) *Conversely, if τ is a Gaussian prime, then τ is an associate of a rational prime congruent to 3 modulo 4, or else $N(\tau) = 2$ or $N(\tau)$ is a rational prime congruent to 1 modulo 4.*

PROOF. (i) Suppose p is a rational prime. Assume $p \equiv 3 \pmod 4$. If p is not prime in **G**, then there are non-unit Gaussian integers α and β such that $\alpha\beta = p$. Thus

$$N(\alpha)N(\beta) = N(p) = p^2,$$

and consequently,

$$N(\alpha) > 1 \quad \text{and} \quad N(\beta) > 1 \quad \text{imply} \quad N(\alpha) = N(\beta) = p.$$

If $\alpha = a + bi$, then a and b are in **Z** and

$$a^2 + b^2 = N(\alpha) = p.$$

But we have

$$a^2 + b^2 \equiv 3 \pmod 4.$$

This is impossible since the squares modulo 4 are 0 and 1.

We now consider the remaining cases. If $p = 2$, then an application of $N(1+i) = 2$ and Theorem 6.3.5 yield that $1+i$ is prime since 2 is a rational prime. If $p \equiv 1 \pmod 4$, then we have an integer $a \in \mathbf{Z}$ such that

$$a^2 + 1 \equiv \pmod p.$$

Hence there is $b \in \mathbf{Z}$ such that

$$a^2 + 1 = pb \quad \text{or} \quad (a+i)(a-i) = pb.$$

If p were a Gaussian prime, then since $p|(a+i)(a-i)$ we would have

$$p|(a+i) \quad \text{or} \quad p|(a-i).$$

However, this is also impossible since

$$\text{neither}\quad a/p + i/p \quad\text{nor}\quad a/p - i/p$$

is a Gaussian integer. Accordingly, p is not a prime in $\mathbf{G}$, and so there are non-unit Gaussian integers τ_1 and τ_2 such that $p = \tau_1\tau_2$ and so $N(\tau_1)N(\tau_2) = p^2$. Owing to $N(\tau_1) > 1$ and $N(\tau_2) > 1$, we have $N(\tau_1) = N(\tau_2) = p$, whence deriving that both τ_1 and τ_2 are Gaussian primes due to Theorem 6.3.5.

(ii) Let τ be a prime in $\mathbf{G}$. Then $N(\tau) \in \mathbf{N}$ and is greater than 1, and is then a product of rational primes:

$$\tau\bar{\tau} = N(\tau) = p_1 p_2 \cdots p_n,$$

where p_j are rational primes. Thus $\tau | p_1 p_2 \cdots p_n$ and hence $\tau | p_j$ for some j.

If $p_j \equiv 3 \,(\mathrm{mod}\, 4)$, then p_j is a prime in $\mathbf{G}$ and thus p_j/τ is a unit and τ is an associate of p_j. Otherwise, if $p_j \not\equiv 3 \,(\mathrm{mod}\, 4)$, equivalently, $p_j = 2$ or $p_j \equiv 1 \,(\mathrm{mod}\, 4)$, then p_j is not a prime and as we have seen earlier, $N(\tau') = p_j$ for some Gaussian prime τ'. Since $\tau' | p_j$, $\tau' | \tau$ or $\bar{\tau}$ and hence τ' is an associate of τ or $\bar{\tau}$. In any case $N(\tau) = p_j$. This completes the argument. □

Remark 6.3.6. From Theorem 6.3.6 we can say that the Gaussian primes comprise $1 + i$, all rational primes congruent to 3 modulo 4, $a \pm bi$ where $a^2 + b^2$ is a rational prime $p \equiv 1 \,(\mathrm{mod} 4)$, and all associates of these primes.

Example 6.3.2. Here are Gaussian prime factorizations of the first ten natural numbers:

$$\begin{aligned}
1 &= 1,\\
2 &= -i(i+1)^2, \quad 3 = 3,\\
4 &= -(i+1)^4, \quad 5 = -i(2i+1)(2+i), \quad 6 = -3i(i+1)^2,\\
7 &= 7, \quad 8 = i(i+1)^6, \quad 9 = 3^2, \quad 10 = -(i+1)^2(2i+1)(i+2).
\end{aligned}$$

Exercises

6.1. Suppose it was enciphered with a Caesar with shift constant k, decipher the message UWJUF WJYTR JJYYM DITTR.

6.2. Encipher the message:

FOUR SCORE AND SEVEN YEARS AGO

using a Caesar cipher with $k = 20$.

6.3. Suppose $p = 11$, $q = 3$ and $j = 3$ in RSA algorithm. Find their encryption and decryption functions.

6.4. Let Jay's choice of p, q, j be 7, 11, 7 respectively. If Alison wishes to send as a message to Jay the single letter Z (whose integer value is 25), what ciphertext does she actually send, and how does Jay decipher it?

6.5. Verify that $(3, 4, 5)$ is the only primitive Pythagorean triple consisting of consecutive integers.

6.6. Prove that $x^2 + y^2 = z^2$ and $(x, y, z) = 1$ yield $(x, y) = (y, z) = 1$.

6.7. Prove that if x is even or odd then $x^4 \equiv 0$ or $1 \pmod{16}$, and then verify that any solution to $x \equiv 15 \pmod{16}$ cannot be written as a sum of fewer than 15 fourth powers.

6.8. Complete the argument for Theorem 6.2.3 by handling the case $(2a, a^2 + 3b^2) = 3$.

6.9. Determine whether the following divisibilities hold:

(i) $(2 + 3i)|(5 - i)$;
(ii) $3|(2 - i)(3 + i)$;
(iii) $(3 - 2i)|26$;
(iv) $(4 - 5i)|(91 - 22i)$;
(v) $(1 + 2i)|41$.

6.10. Let $\alpha, \beta \in \mathbf{G}$. Show that if $\alpha|\beta$ then $N(\alpha)|N(\beta)$. Does the converse hold?

6.11. Prove that $(1 + i)|(x + iy)$ iff $x \equiv y \pmod 2$, and then verify that $\alpha \in \mathbf{G}$ implies $\alpha \equiv 0$ or $1 \pmod{(1 + i)}$.

6.12. Factorize $18 + 246i$ into Gaussian primes.

Solutions to Even-numbered Exercises

1. Basics

Ex. 1.2. Clearly the formula is true for $n = 1$. Let it be true for $n = k$. If $n = k + 1$, then

$$\begin{aligned}\sum_{i=1}^{k+1} \frac{1}{i^2+i} &= \sum_{i=1}^{k} \frac{1}{i^2+i} + \frac{1}{(k+1)(k+2)} \\ &= \frac{k}{k+1} + \frac{1}{(k+1)(k+2)} \\ &= \frac{k+1}{k+2}\end{aligned}$$

and hence by I_1 the formula is true for all $n \in \mathbf{N}$.

Ex. 1.4. Clearly the formula is true for $n = 1$. Let it be true for $n = k$. If $n = k + 1$, then

$$\begin{aligned}\sum_{i=1}^{k+1} i^3 &= \sum_{i=1}^{k} i^3 + \frac{1}{(k+1)^3} \\ &= \left(\frac{(k+1)(k+2)}{2}\right)^2\end{aligned}$$

and hence by I_1 the formula is true for all $n \in \mathbf{N}$.

Ex. 1.6. Clearly the formula is true for $n = 1$. Let it be true for $n = k$. If $n = k + 1$, then

$$\begin{aligned}x^{2(k+1)} &- x^{2(k+1)-1} + \cdots + x^2 - x + 1 \\ &= x^{2(k+1)} - x^{2k+1} + \frac{x^{2k+1}+1}{x+1} \\ &= \frac{x^{2(k+1)}+1}{x+1}\end{aligned}$$

and hence by I_1 the formula is true for all $n \in \mathbf{N}$.

Ex. 1.8. Suppose $\alpha = (1 + \sqrt{5})/2$ and

$$\mathcal{F}_n = \left(\left(\frac{1+\sqrt{5}}{2}\right)^n - \left(\frac{1-\sqrt{5}}{2}\right)^n\right) \Big/ \sqrt{5}.$$

Clearly $\mathcal{F}_3 = 2 > \alpha$ and $\mathcal{F}_4 = 3 > \alpha^2$, and so the inequality is true for $n = 3, 4$. Let it be true for $n = k, k+1$, $k \geq 3$. If $n = k+2$, then

$$\mathcal{F}_{k+2} = \mathcal{F}_{k+1} + \mathcal{F}_k > \alpha^{k-1} + \alpha^{k-2} = \alpha^k$$

due to $\alpha + 1 = \alpha^2$. Thus, the inequality is true for all $n \geq 3$ by I_4.

Ex. 1.10. Suppose that $\sqrt{5}$ is rational. So there exists a least positive integer b and $a \in \mathbf{N}$ such that $\sqrt{5} = a/b$. Consequently, $2 < a/b < 3$, i.e., $0 < a - 2b < b$. Since $\sqrt{5} = a/b$, it follows that

$$5b^2 = a^2, \quad b(5b - 2a) = a(a - 2b) \quad \text{and} \quad \sqrt{5} = \frac{a}{b} = \frac{5b - 2a}{a - 2b}.$$

But this gives $\sqrt{5}$ as the ratio of the two integers where the denominator is positive but less than b. Since this is a contradiction, $\sqrt{5}$ is irrational.

Ex. 1.12. By the division algorithm, there exist integers u and r with $0 \leq r < |b|$ such that $a = |b|u + r$. Since $b < 0$, one has $|b| = -b$. Therefore, setting $-u = q$, we obtain

$$a = -bu + r = bq + r,$$

as desired.

2. Divisibility

Ex. 2.2. If $r \neq 0$ and $a = bq + r$, then $r = a - bq$. Since $b|a$, it follows that $b|r$ and that there is $t \in \mathbf{Z}$ such that $r = bt$ and so $|b| \leq r$. This contradicts $0 \leq r < b$.

Ex. 2.4. $(357, 629) = 17 = 357(-7) + 629 \cdot 4$.

Ex. 2.6. If $b|a$, then there is $q \in \mathbf{Z}$ such that $bq = a$. Thus, $a = \pm q|b|$ and $|b|$ divides a. Since $|b|$ divides b, $|b|$ is a common divisor of a and b. If d is any other common divisor, then $d|a, b$, and hence d divides $|b|$. Thus, $(a, b) = |b|$. Conversely, if $(a, b) = |b|$, then it follows that $|b|$ divides a so that $a = |b|q$ for some $q \in \mathbf{Z}$. But then $a = \pm qb$ and $b|a$ as claimed.

Ex. 2.8. Let $d = (a + b, a - b)$. Since

$$2a = a + b + (a - b) \quad \text{and} \quad 2b = a + b - (a - b),$$

it follows that d divides $2a$ and $2b$. Therefore, $d|(2a, 2b)$. However $(2a, 2b) = 2(a, b) = 2$. Thus $d|2$ and so $d = 1$ or $d = 2$.

Ex. 2.10.

$$[357, 629] = 357 \cdot 629/17 = 13209$$

and

$$[1492, 1066] = 1492 \cdot 1066/2 = 795236.$$

Ex. 2.12. Since the case of the *lcm* is exactly analogous to that of the *gcd*, one considers only the *gcd*. The *gcd* of $k \geq 2$ nonzero integers a_i, $i = 1, 2, \ldots, k$ is the unique positive $g = (a_1, a_2, \ldots, a_k)$ with the following two properties:

(i) $g|a_i$ for each i;
(ii) If $d|a_i$ for each i, then $d|g$.

Now let us prove

$$g = ((a_1, a_2, \ldots, a_{k-1}), a_k).$$

In fact, let

$$h = \big((a_1, a_2, \ldots, a_{k-1}), a_k\big).$$

Then g and h are positive. Since $g|a_i$, we have $g|(a_1, a_2, \ldots, a_{k-1})$ and $g|a_k$, whence deriving $g|h$. On the other hand, $h|a_k$ and $h|(a_1, \ldots, a_{k-1})$ imply $h|a_j$ for each i and so $h|g$. Therefore $g = h$.

3. Primes

Ex. 3.2. Since $3, 7$ and 11 are of this form, the following argument is not vacuous. Assume that there are only finitely many primes of this form, say $p_1, p_2, \ldots, p_r$ and consider the number $m = 4p_1p_2\cdots p_r - 1$. Since m is of the form $4k-1$ and $m > p_i$ for each i, it follows that m is composite and must have prime factors of the form $4k+1$ or $4k-1$. Since the product of any two numbers of the form $4k+1$ is again of that form, it follows that m has at least one prime divisor of the form $4k-1$. Thus, $p_i|m$ for some i. However, then $p_1|1$, and this is impossible. Therefore, there must be infinitely many primes of the form $4k-1$.

Ex. 3.4. Since

$$4725 = 3^3 \cdot 5^2 \cdot 7^1 \quad \text{and} \quad 3234 = 2^1 \cdot 3^1 \cdot 7^2 \cdot 11,$$

we conclude

$$(4725, 3234) = 2^0 \cdot 3^1 \cdot 5^0 \cdot 7^1 \cdot 11^0 = 21$$

and

$$[4725, 3234] = 2^1 \cdot 3^3 \cdot 5^2 \cdot 7^2 \cdot 11^1 = 727650.$$

Ex. 3.6. Let $a = \prod_{i=1}^r p_i^{k_i}$ with $k_i > 0$ for each i. Then the number of positive divisors of a is $\tau(a) = \prod_{i=1}^r (k_i + 1)$ and hence $\tau(a)$ is odd iff $k_i + 1$ is odd for all i, that is, iff k_i is even for each i. However, it is so iff a is a square by the standard factorization theorem.

Ex. 3.8. $[2.7 + 0.5] = 3$, $[-75/4] = -19$, $[-\sqrt{2} + 0.5] = -1$.

Ex. 3.10. (i) If $a \in \mathbf{Z}$ is not $[n + 1/2]$, then $|a - [n+1/2]| \geq 1$ and hence

$$|a - n| \geq |a - [n + 1/2]| - |[n + 1/2] - n| \geq 1/2$$

if we verify that

$$|[n + 1/2] - n| \leq 1/2.$$

To do so, note that

$$n - 1/2 < [n + 1/2] \leq n + 1/2.$$

Thus we have

$$-1/2 < [n + 1/2] - n \leq 1/2,$$

giving the desired inequality.

(ii) If $n \in \mathbf{Z}$ then $[n+1/2] = n$ since in this case $n-1$ and $n+1$ are two integers equally near n but $n+1 \neq [n+1/2]$. Meanwhile, if $n = k+1/2 \in \mathbf{R} \setminus \mathbf{Z}$ where $k \in \mathbf{Z}$, then $a = k-1$ and $b = k+2$ give $n - a = 1.5 = b - n$ with

$$[n+1/2] = k+1 < k+2 = b,$$

i.e., $[n+1/2]$ is not greater than b.

Ex. 3.12. $a^n - 1 = (a+1)(a^{n-1} - a^{n-2} + a^{n-3} - \cdots + a - 1)$. If $a = 1$ then $a^n + 1$ has $a+1$ as a factor, but if $a > 1$ then $a+1$ is not a factor of $a^n + 1$. If, otherwise, $a+1$ is a factor, then $(a+1)|a^n+1$. But noting that n is even and

$$a^n = (a+1-1)^n = (a+1)^n - n(a+1)^{n-1} + \cdots - n(a+1) + 1,$$

we obtain that $(a+1)|2$, contradicting $a > 1$.

Ex. 3.14. Suppose $g = (a, b_1 b_2 \cdots b_n) > 1$. Then there is a prime p such that $p|g$. Since $g|a$ and $g|b_1 b_2 \cdots b_n$, it follows that $p|a$ and $p|b_1 b_2 \cdots b_n$. Thus $p|b_i$ for some i, which contradicts $(p, b_i) = 1$ for each i. Therefore, it must be the case that $g = 1$.

Ex. 3.16. It is clear that $\prod_{i=1}^{r}(1+k_i)$ is the number of positive divisors of $a = \prod_{i=1}^{r} p_i^{k_i}$ and that the sum $\sigma(a)$ of these divisors is given by

$$\prod_{i=1}^{r}(1 + p_i + p_i^2 + \cdots + p_i^{k_i}) = \prod_{i=1}^{r} \frac{p_i^{k_i} - 1}{p_i - 1},$$

as claimed.

Ex. 3.18. By L'Hospital's Rule, one has

$$\lim_{x\to\infty} \frac{\mathrm{li}(x)}{x/\ln x} = \lim_{x\to\infty} \frac{\ln x}{\ln x - 1} = 1.$$

Thus the formula:

$$\frac{\mathrm{li}(x)}{x/\ln x} = \frac{\mathrm{li}(x)}{\pi(x)} \cdot \frac{\pi(x)}{x/\ln x}$$

implies the desired result.

4. Congruences

Ex. 4.2. Note that

$$\begin{aligned} 10 &\equiv 3 \pmod 7 \\ 10^2 &\equiv 3^2 \equiv 2 \pmod 7 \\ 10^3 &\equiv 20 \equiv -1 \pmod 7). \end{aligned}$$

So it follows that

$$10^{515} \equiv 10^{3\cdot 171+2} = (10^3)^{171} 10^2 \equiv (-1)^{171} \cdot 2 \equiv 5 \pmod 7).$$

Ex. 4.4. Modulo 3 any integer a is congruent to one of -1, 0, or 1. Consequently, a^3 is congruent to -1, 0, or 1, respectively, and $a^3 \equiv a \pmod 3$ in every case. Multiplying by a^2, we have

$$a^5 \equiv a^3 \pmod 3 \quad \text{and thus} \quad a^5 \equiv a \pmod 3.$$

Similarly, any a is congruent modulo 5 to one of $-2, -1, 0, 1$ or 2. Since $2^5 = 32 \equiv 2 \pmod 5$, it follows that $(-2)^5 \equiv -2 \pmod 5$. The computations are trivial in the other cases, so again $a^5 \equiv a \pmod 5$ in every case.

Ex. 4.6. Because $\phi(7) = 6$ and because all the given numbers are relatively prime to 7:

$$(-30, 7) = 1,\ (-19, 7) = 1, \text{ and so on,}$$

and no two numbers out of them are congruent to each other modulo 7, the given numbers form a reduced system modulo 7.

Ex. 4.8. Since $\phi(p) = p - 1$, a reduced residue system modulo p must contain $p-1$ integers, each relatively prime to p and no two of which are congruent modulo p. Thus, we have only to show that $(a^i, p) = 1$ for $i = 1, 2, \ldots, p-1$ and that $a^i \not\equiv a^j \pmod p$ for $i \neq j$. The first of these results is clearly true since $(p, a) = 1$. Now suppose

$$a^i \equiv a^j \pmod p \quad \text{with} \quad 1 \le i < j \le p-1.$$

Because $(a^j, p) = 1$ for each j, we may divide both sides of this congruence by a^j to obtain

$$a^{i-j} \equiv \pmod p \quad \text{and} \quad 1 \le i - j \le p - 2.$$

But this violates the condition that $p-1$ is the smallest positive value of k such that $a^k \equiv 1 \pmod p$. Therefore, $a^i \not\equiv a^j \pmod p$ for $i \neq j$ and the proof is complete.

Ex. 4.10. By Fermat's Little Theorem, we have

$$a^p \equiv a \pmod p \quad \text{and} \quad b^p \equiv b \pmod p.$$

Since $a^p \equiv b^p \pmod p$, it follows that $a \equiv b \pmod p$. Accordingly, $a = b + pr$ for some integer r and, by the binomial theorem,

$$\begin{aligned} a^p &= (b+pr)^p \\ &= b^p + pb^{p-1}(pr) + \frac{p(p-1)}{2} b^{p-2}(pr)^2 + \cdots + (pr)^p \\ &= b^p + p^2 q, \end{aligned}$$

where q is an integer since all terms on the right factor the first are integers divisible by p^2. Therefore, $a^p \equiv b^p \pmod{p^2}$, as was to be shown.

Ex. 4.12. Since $\phi(14) = 6$, every integer which is relatively prime to 14 and has order modulo 14 is a divisor of 6. Clearly, 1 has order 1 modulo 14. Since

$$3^2 \equiv 9 \pmod{14} \quad \text{and} \quad 3^3 \equiv 27 \pmod{14},$$

we see that 3 cannot have order $1, 2$ or 3; hence it must have order 6. The following table gives the orders modulo 14 of all the elements of the reduced residue system exhibited above:

Integer	1	3	5	-5	-3	-1
Order	1	6	6	3	3	2

Ex. 4.14. If $s \in \mathbf{N}$ then $(a^k)^s \equiv a^{ks} \equiv 1 \pmod{m}$ if and only if $r|ks$, since a has order r modulo m. Set $g = (k, r)$. Then $r = r_1 g$ and $k = k_1 g$ for some $r_1, k_1 \in \mathbf{N}$. Hence $r|ks$ if and only if $r_1|k_1 s$. However, $(r_1, k_1) = 1$, so $r_1|k_1 s$ if and only if $r_1|s$. We have thus shown that $(a^k)^s \equiv 1 \pmod{m}$ if and only if $r_1|s$. Clearly, the smallest such positive integer s is r_1, and a^k has order $r_1 = r/(k, r)$ modulo m.

5. Congruent Equations

Ex. 5.2. (i) $x = 5$; (ii) $x = 45, 94$.

Ex. 5.4. $x_k = 5 + 7k$, $y_k = 5 - 15k$ for $k \in \mathbf{Z}$.

Ex. 5.6. The solution 23 is unique modulo 105.

Ex. 5.8. Solutions of

$$3x \equiv 7 \pmod{5},\ x \equiv 1 \pmod{4},\ 5x \equiv 2 \pmod{11}$$

are, respectively, $-1, 1$, and -4. Using CRT to solve simultaneously the system:

$$x \equiv -1 \pmod{5},\ x \equiv 1 \pmod{4},\ x \equiv -4 \pmod{11},$$

we obtain 29 as a solution to this system and hence to the original one. The solution is unique modulo 220.

Ex. 5.10. As indicated in the hint in the statement of the problem,

$$p - 1 \equiv -1 \pmod{p},\ p - 2 \equiv -2 \pmod{p},\ \ldots,\ p - k \equiv -k \pmod{p}.$$

Therefore,

$$(p-1)(p-2)\cdots(p-k) \equiv (-1)^k k! \pmod{p}.$$

Also, since $h + k = p - 1$, it follows that $h = p - k - 1$ and $h! = (p - k - 1)!$. Combining this with the above equality involving $k!$, we now obtain

$$(p-1)! \equiv (-1)^k h! k! \pmod{p}$$

or, on multiplying trough by $(-1)^k$,

$$h!k! \equiv (-1)^k (p-1)! \equiv (-1)^{k+1} \pmod{p}$$

since $(p-1)! \equiv -1 \pmod{p}$, by Wilson's Theorem.

Ex. 5.12. Using the table of indices to base 2 modulo 11, we have

$$x^{25} \equiv 3 \pmod{11} \quad \text{iff} \quad 25\text{ind}_2 x^{25} \equiv 3 \pmod{10}$$
$$\text{iff} \quad 25\text{ind}_2 x \equiv 8 \pmod{10}.$$

Since $(25, 10) \nmid 8$, this congruence is not solvable.

Ex. 5.14. If $M = \sum_{i=1}^{(p-1)/2}[i2/p]$, then $M = 0$ and hence

$$(p^2-1)/8 \equiv n \pmod 2$$

by

$$P(q-1) = (M+n)p + 2\sum_{i=1}^{n} a_i,$$

where n is the number of elements of $\{q, 2q, \ldots, (p-1)/2q\}$ which are congruent modulo 2 to negative elements of

$$\{-(p-1)/2, \ldots, -2, -1, 1, 2, \ldots, (p-1)/2\}.$$

Since $(2/p) = (-1)^n$, we have $(2/p) = (-1)^{(p^2-1)/8}$. This result, together with Theorem 5.3.2 (ii) (iv) (v) and Theorem 5.3.4, implies that $(-23/59) = 1$ and so that -23 is a quadratic residue of 59.

Ex. 5.16. We get a series of congruences all modulo 17 below:

$$\begin{aligned} 3x^2+x+3 &\equiv 0, \\ 18x^2+6x+18 &\equiv 0, \\ x^2+6x &\equiv -1, \\ x^2+6x+9 &\equiv 8, \\ (x+3)^2 &\equiv 25, \\ x+3 &\equiv \pm 5, \\ x &\equiv 2,\ -8. \end{aligned}$$

6. Three Additional Topics

Ex. 6.2. ZIOLM WILYU HXMYP YHSYU LMUAI.

Ex. 6.4. Since $j = 7$, $N = 25$, $M = 53$ and $53 \equiv 25^7 \pmod{77}$, Alison transmits $M = 53$ to Jay. From $j = 7$ it follows that $k = 43$ will have been previously calculated by Jay to ensure $43j \equiv 1\,(\mathrm{mod}, 60)$. Now Jay evaluates $25 \equiv 53^{43} \pmod{77}$ and recovers the plaintex message Z.

Ex. 6.6. Assume that $(x, y) = d > 1$. Then there is a prime p such that $p|d$. Since $d|x, y$, it follows that $p|x, y$, and hence also that $p|x^2, y^2$. But then $p|z^2$ due to $x^2 + y^2 = z^2$. But then $p|z$ and this violates $(x, y, z) = 1$. Similarly, if $(y, z) = d > 1$, then there exists a prime p such that $p|d$. Since $d|y, z$, it follows that $p|y, z$ and hence $p|y^2, x^2$. But then $p|x^2$ since $x^2 = z^2 - y^2$, and it follows that $p|x$. This violates $(x, y, z) = 1$ and the proof is complete.

Ex. 6.8. Let $(2a, a^2 + 3b^2) = 3$. In this case, we have $a = 3c$ where c is even and $3 \not| b$, whence deriving

$$z^3 = 6c(9c^2 + 3b^2) = 18c(3c^2 + b^2),$$

where $(18c, 3c^2 + b^2) = 1$. As a matter of fact, since c is even and b is odd, we get that $3c^2 + b^2$ is odd, and $3 \not| (3c^2 + b^2)$ and $(b, c) = 1$. The standard factorization of integers into primes allows us to choose r which is divisible by 3, and s which is odd, such that

$$18c = r^3 \quad \text{and} \quad 3c^2 + b^2 = s^3.$$

Once again, we can write $s = u^2 + 3v^2$, $u, v \in \mathbf{Z}$ and

$$b = u(u^2 - 3v^2), \quad c = 3v(u^2 - v^2).$$

Clearly, u is odd, v is even, $v \neq 0$ and $(u, v) = 1$, but also $2v$, $u + v$, and $u - v$ are pairwise relatively prime. This done, we may obtain

$$r^3 = 18c = 54v(u - v)(u + v),$$

and so $2v$, $u + v$ and $u - v$ are cubes:

$$2v = \gamma, \quad u + v = \beta^3, \quad u - v = -\alpha^3,$$

where $\alpha, \beta \in \mathbf{Z} \setminus \{0\}$ and γ is even but not zero. Nevertheless,

$$|z|^3 = 18|c|(3c^2 + b^2) = 27|\gamma|^3|u^2 - v^2|(3c^2 + b^2) > |\gamma|^3$$

contradicts the minimality of $|z|$.

Ex. 6.10. Since $\alpha|\beta$, there is a $\gamma \in \mathbf{G}$ such that $\beta = \gamma\alpha$. This yields

$$N(\beta) = N(\gamma\alpha) = N(\gamma)N(\alpha) \quad \text{and so} \quad N(\alpha)|N(\beta).$$

Conversely, it is not true. Below is the counterexample: $\alpha = 3 + 4i$, $\beta = 4 + 3i$. It is clear that $N(\alpha) = N(\beta)$ but $\alpha \nmid \beta$ and $\beta \nmid \alpha$.

Ex. 6.12. Since $18 + 246i = 6(3 + 41i)$, it suffices to factorize $3 + 41i$. To the end, note that

$$(3 + 41i)(3 - 41i) = N(3 + 41i) = 3^2 + 41^2 = 2 \cdot 5 \cdot 13^2.$$

So the Gaussian prime factors of $3 + 41i$ are among the prime factors of 2, 5, 13, i.e., $1 + i$, $2 \pm 3i$, $1 \pm 2i$ thanks to $5, 13 \equiv 1 \pmod 4$. By checking these five Gaussian primes, we get

$$\begin{aligned} 3 + 41i &= (2 - 3i)(-9 + 7i) \\ &= (2 - 3i)^2(-3 - i) \\ &= (2 - 3i)^2(1 + i)(-2 + i), \end{aligned}$$

whence getting

$$18 + 246i = 3(1 + 2i)(2 - 3i)^2(1 + i)^3.$$

Bibliography

[1] G.A. Jones and J.M. Jones, *Elementary Number Theory*, SUMS, Springer, 1998.

[2] C.T. Long, *Elementary Introduction to Number Theory*, Prentice-Hall, Inc., Englewood Cliffs, New Jersey, 1987.

[3] N.H. McCoy, *The Theory of Numbers*, The Macmillan Company, Collier-Macmillan Limited, London, 1969.

[4] I. Niven, H.S. Zuckerman, and H.L. Montgomery, *An Introduction to the Theory of Numbers*, fifth edition, John Wiley & Sons, Inc., New York, 1991.

[5] P. Ribenboim, *Fermat's Last Theorem for Amateurs*, Springer-Verlag, New York, Inc., 1999.

[6] D.E. Rideout, *Pure Mathematics,* **3370**: *Course Notes*, Memorial University of Newfoundland, 2001.

Index

Made in United States
Orlando, FL
05 November 2026

90230228R00057